Introduction to Clay Minerals

Introduction to Clay Minerals

Chemistry, origins, uses and environmental significance

B. Velde

Director of Research, National Centre for Scientific Research, France.

CHAPMAN & HALL
London · Glasgow · New York · Tokyo · Melbourne · Madras

Published by Chapman & Hall, 2–6 Boundary Row, London SE1 8HN

Chapman & Hall, 2–6 Boundary Row, London SE1 8HN, UK

Blackie Academic & Professional, Wester Cleddens Road, Bishopbriggs, Glasgow G64 2NZ, UK

Chapman & Hall, 29 West 35th Street, New York NY10001, USA

Chapman & Hall Japan, Thomson Publishing Japan, Hirakawacho Nemoto Building 6F, 1-7-11 Hirakawa-cho, Chiyoda-ku, Tokyo 102, Japan

Chapman & Hall Australia, Thomas Nelson Australia, 102 Dodds Street, South Melbourne, Victoria 3205, Australia

Chapman & Hall India, R. Seshadri, 32 Second Main Road, CIT East, Madras 600 035, India

First edition 1992

Typeset in 10/12 pt by Excel Typesetter Company, Hong Kong
Printed in Great Britain by the University Press, Cambridge

ISBN 0 412 37030 1

A catalogue record for this book is available from the British Library

Library of Congress Cataloging-in-Publication data available

Contents

Preface

This is the first page of a rather short, diagram-filled book which has the pretention of informing the unwary reader of the importance of clay minerals in his or her daily life. This message is difficult to grasp without some background in chemistry and a little less in the physics of the common measurement processes, X-ray generation and X-ray interaction, which are essential to the characterization of clay minerals. However, one should not be daunted by the task; it is in fact rather easy to understand 'the meaning of clays'. But why should we? The habitat of clay minerals is essentially that of man, the very thin surface of contact between air, water and earth. The clays belong to the earth but owe most of their characteristics to the interaction of water and air. So does man. We are thus intimately related to clays, more that the biblical prophecy of origin and destiny might lead us to believe. For in fact, man's behaviour in the clay contact zone affects all of his fellow living creatures, whether he wishes it or not. Clays are a filter and a substrate for life. They reflect the chemistry of the surface. One must understand their origins and their characteristics to predict their behaviour. It is hoped that in this short work one can glimpse the importance and grasp the fundamentals of clay science.

As in all prefaces, the author wishes to recognize the fact that he did not invent the study of the material he discusses. It is in fact more human than that, for without the inspiration of John Hower the author would not have begun nor thought about such an undertaking.

B. Velde

1

The clay perspective

1.1 WHAT ARE CLAY MINERALS?

Clay minerals are the fine-grained part of geology. Clays were initially defined as consisting of grains less than 2 μm in diameter, beyond the limit of microscopic resolution. This was a definition of the nineteenth century, which had only microscopic means of investigation.

Of the mineral types or families designated by virtue of their grain size as clay, many also have a mineral structure in common. Most clays in natural settings have a **phyllosilicate** or sheet structure. Such a structure implies that the ratio of particle dimensions is somewhat akin to that of a sheet of paper, that is to say, much larger and longer than thick. Other minerals of different grain shapes present in the clay fraction include zeolites, quartz and oxide minerals as well as minor portions of all of the other minerals found in the geological environment. However, most geologists mean sheet silicates when they speak of clay minerals.

Statistically, geologists are correct in their nomenclature, clay minerals have a common structure but they are almost always the result of chemical changes or thermal variations in the range of near-surface conditions. The maximum range of temperatures at which clay minerals occur can be given as 4–250°C. At higher temperatures, the sheet silicate minerals are considered to be of metamorphic origin and they have a strong tendency to have a larger grain size than that given to indicate the clay mineral realm. They do not have the same composition as minerals of lower temperature origin.

1.1.1 Basic nomenclature

Several basic properties of clay minerals guide all of the study devoted to them. The most important is the capacity of certain clays to change volume by absorbing water molecules or other polar ions into their structure. This is called the swelling property. Clays are thus divided into **swelling** and **non-swelling** type minerals. Swelling clays are called

smectites. Small particles of other non-sheet silicate materials can have some of the properties of clays, forming gels or thixotropic states, but the swelling properties are unique to the clay mineral world.

The next important property is the basic composition and structure of the clays which is used to classify the remaining clay minerals. Two clay mineral groups are similar to metamorphic minerals and they are usually compared to these higher-temperature phases. One type is the potassic, mica-like minerals, which are dominated by **illite**. Also one finds **chlorites** which are compositionally and structurally contiguous with the high-temperature chlorite phases.

Of those clay types which remain, there is the **kaolinite** family which has no equivalent in metamorphic mineral groups and the needle-shaped **sepiolite-palygorskites**, which are found uniquely in low-temperature environments. The mineral groups smectite, illite, chlorite, kaolinite, sepiolite-palygorskite account for by far the greatest part of the clay minerals found in nature.

1.1.2 Clay properties

The key to clay mineralogy is the small diameter of the mineral grains, less than 2 μm, and their crystallographic habit, which is sheet-like. These two factors give clays a very high surface area relative to the mass of material in the clay mineral crystal grains. This large surface area gives the properties of water adsorption common to all fine-grained materials. The surface residual charges on the mineral structure attract and adsorb water in layers, up to four perhaps. These layers are structured in a loose fashion but do not leave the surface easily. Figure 1.1 shows the increase in depth and, hence, lithostatic pressure necessary to desorb these water layers on clays.

Clays, then, attract water to their surfaces. This attraction creates a reservoir of water in their environment. Their adsorbed water skin gives clays a greater tendency to remain in suspension in aqueous media and hence they can be easily transported in this suspended state.

All clays attract water to their surfaces (adsorption), but some of them bring it into their structure (absorption). Absorption is the incorporation of molecules into the crystal grain while adsorption is the addition of molecules onto the surface of the grains. The sepiolite-palygorskites contain water in channels in their structure which is easily released upon heating. This is called **zeolitic** water. The smectites attract water between the sheet layers of the structure. In doing so they change their volume greatly. This swelling property is very remarkable in their natural setting, especially just after a rainstorm in a semi-desert area such as the western United States. The dusty, fine-grained soil becomes a sticky, inflated mass

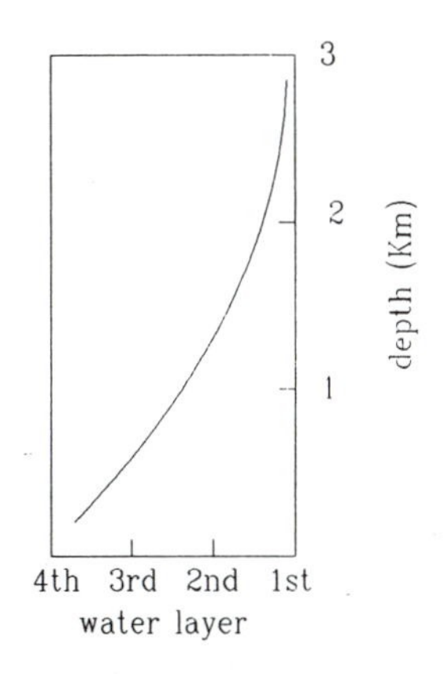

Fig. 1.1 Water desorption (layers of adsorbed water molecules) from clays as a function of burial depth in sediments. The number of adsorbed water layers decreases as the temperature and pressure of the sediments increase.

which adheres to the shoes and vehicle tyres, much to the dismay of an unwary traveller.

We can thus classify clays by the way they absorb water (Fig. 1.2).

1.2 CLAY ENVIRONMENTS AND THE CLAY CYCLE

Clays are an integral part of the geological environment. They are extremely important to that environment which reaches the activities of man. They are born and disappear in a cycle which can be described as follows.

The origin of the greatest amount of clay material is in the process of **weathering**, either subaerial or subaquatic. Sedimentation and burial changes the clay species. Clays are transformed into other clays. Some clay material is produced by hydrothermal processes (water–rock interaction at temperatures of 100–250°C). Hydrothermal alteration gives us a large proportion of the high-quality clays used in industrial processes.

The mutation or transformation of clays from their origins in weathering, through their transportation by fluvial means to sedimentary basins where they are subsequently buried, is a major part of what can be called the clay cycle. At each step the clays respond to their chemical and thermal environment and their properties and species change. We can look briefly at these environments.

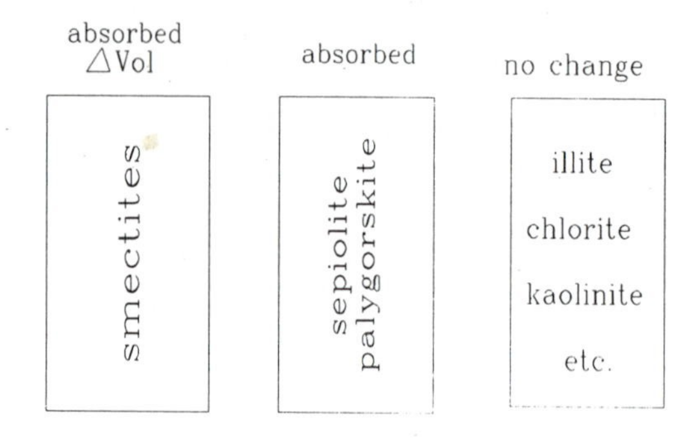

Fig. 1.2 Clay mineral families according to water sorption characteristics. Smectites are swelling clays, where water absorption increases the volume of the mineral. Sepiolite-palygorskites do not swell but have an important sorption capacity. Other clay minerals have neither of these characteristics.

1.2.1 Weathering

At the rock–atmosphere (or earth–air) interface, high-temperature silicate minerals become unstable due to changes in the chemistry of their environment. The new chemical conditions are dominated by the aqueous state. Here the ratio of water to rock is very great. The consequence of such an environment is a tendency to hydrate the high-temperature silicate minerals. This hydration is effected during a structural transfer which releases a portion of the old mineral into solution and hydrates the more insoluble residue. The resulting mineral structures contain crystalline water, that is, hydrogen bonded in the form of OH units in the interior of the mineral (Fig. 1.3).

Smectites and sepiolite-palygorskites also contain absorbed molecular water (H_2O) which is loosely held. Therefore the hydrolysis process which occurs during weathering involves two kinds of hydrogen, bound as either OH or H_2O, which is found in two different types of crystallographic site. Under laboratory or factory conditions the crystalline water is lost at higher temperatures (above 500°C) than the absorbed interlayer and zeolitic water (below 120°C).

Even though a large portion of clays are produced by weathering, the major effect during this process is in fact that of **dissolution**. A very large portion of the mineral material in a rock is dissolved integrally into the altering aqueous solution (rainwater) and it is transported as such into the collecting basins of lakes or the ocean. Only a proportion of the initial rock-forming silicate minerals is dissolved incongruently (i.e. change of a mineral phase into another mineral phase and some ions dissolved in aqueous solution), leaving behind clay minerals (sheet silicates) and oxides

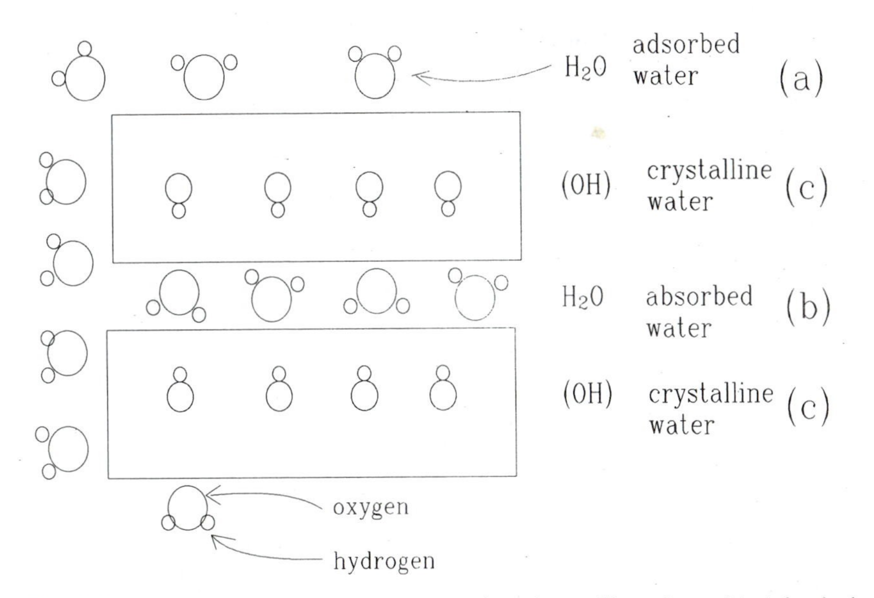

Fig. 1.3 Different types of water associated with swelling clays. (a) Adsorbed water molecules fixed on the clay surface. (b) Bound or absorbed water molecules usually associated with cations found between the layers of the clay structure. (c) Crystalline water which is present in the form of OH units, which 'oxidize' upon heating, forming H_2O molecules.

(principally Fe^{3+}). Clays formed during weathering processes are the result of incongruent dissolution of silicates which were initially formed at conditions of higher temperature and pressure.

The clays formed in the weathering process are eventually taken away by the same rainwater through the processes of **erosion**. Transportation of clays is easy because of the small grain size of the clays which allows them to remain in suspension in the liquid. They are easily moved over the long distances involved in river flow.

1.2.2 Sedimentation

Sedimentation of clays occurs for the most part in lakes or in the ocean. In both environments changes in the clay mineralogy occur at the sediment–water interface, that is, just after their deposition.

At the sediment–water interface, chemical conditions change from those of the freshwater environment of weathering. In the initial stages of sedimentary burial the aqueous solution in the sediment becomes more concentrated in dissolved elements. This changes the chemical equilibria from those of the weathering conditions. The elements in solution have a different chemical potential from those of the clay forming process. The residence time of clays in the sedimentary environment is greater than in the weathering process. Because of this the clays have more time to react with their ambient solution and they become dominant in the silicate-solution chemical equilibrium. The mass of clays reacting compared to the mass of the solution is greater than it was in the weathering situation where the flow rate of solutions was so great as to have swamped the effect of the clays on the activities of elements in the solutions.

Several clay mineral types have their origin uniquely in the sedimentary environment. The reactions which form them are clay–clay and oxide–clay, with a certain input of chemical elements from the sedimentary solutions. The new minerals are greatly affected by the oxidation potential of their environment. The importance of organic material which controls the oxidation state of the silicates in the sedimentary environment is evident.

The action of living organisms in sedimentary material at the sediment–water interface is well known. It can occur by physical bioperturbation, burrowing animals, etc., or it can be effected by microbial action which changes the oxidation state of iron, for example, or which stabilizes sulphides instead of sulphates. This chemical action tends to allow some elements to migrate small distances where they are fixed by a microbial process in a sulphate, sulphide, carbonate mineral. These phases can in turn be destroyed and the elements might migrate upwards. Secondary fixation brings the element into another cycle where it moves downward, and so forth. In short, the sediment–solution interface can be one of recycling of certain elements or possibly of all of the sediment where animals are active.

Surface diagenesis is not of great importance as far as the quantity of material is concerned but the several clay minerals produced there can clearly indicate the existence of the sedimentation environment, either under shallow-water conditions or under those of the deep sea. These minerals are useful indicators of palaeoenvironments.

1.2.3 Deep-sea alteration

The interaction of eruptive rocks and seawater in the deep-sea environment has been seen to effect a specific type of weathering or diagenesis. Basalts (the great majority of the rock types found in ocean-floor conditions) have been seen to hydrolyse, forming clays in their surface layers. This action is especially important in the glassy material found in the eruptive rocks. The clay types produced are typical of this environment and can be used as markers of their origin and that of a residence in deep-sea conditions.

1.2.4 Burial diagenesis

When the clays in sediments are subjected to burial, the ratio of water to rock again changes and the clays become even more dominant in the solution-solid chemical equilibria. There are gradual readjustments under these conditions which produce new minerals from metastable minerals found in the sediments. Further burial changes the temperature of the sediments which changes the mineral stabilities which in turn creates new clays in this environment. The reactions here are for the most part clay–clay in nature. They are not only temperature- but also time-dependent due to slow reaction rates.

1.2.5 Metamorphism

Further burial (in high-temperature conditions) brings the clay minerals to a state of recrystallization which is commonly termed **metamorphic**. This describes thermodynamic conditions which depend almost entirely upon the temperature attained and the length of time for which these temperatures have affected the clays. For example, clays in recent sediments and geothermal areas can persist up temperatures above 200°C. However, reactions can take place at 80°C if the temperature has been maintained for periods of greater than 200 million years. Thus a definition of the upper limit of the clay mineral facies also depends upon the time over which the thermal action has occurred. The clay mineral–metamorphic boundary is a dynamic one governed by reaction kinetics.

Essentially the metamorphic conditions coarsen minerals by crystal growth as well as creating new silicate minerals other than sheet silicates. Metamorphic mineral grains are for the most part too big to be called clays, even though they might have similar chemical and structural properties. Furthermore, their crystallographic composition is no longer that of low-temperature, fine-grained material called clays.

1.2.6 Hydrothermal alteration

A somewhat minor occurrence of clay minerals, but an important one, is the alteration of rocks by the action of large quantities of hot water. This is called **hydrothermal alteration**. It is essentially a type of weathering (high water-to-rock ratio) which is accomplished at temperatures above 50°C. This can occur as the result of the latter stages of cooling of an intrusive, magmatic body at depth. This type of action is often accompanied by the deposit of ore minerals. Symmetrically, most often ore mineralization at depth is accompanied by the formation of clay or clay-like minerals. The same process, linked to the emplacement of basaltic

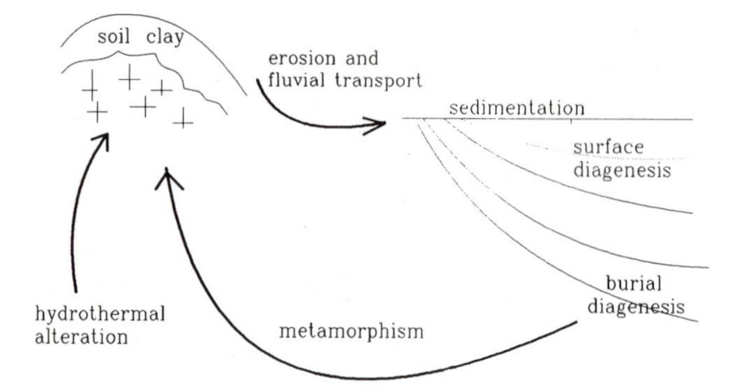

Fig. 1.4 The clay cycle, illustrated as soil formation and erosion, sedimentation after fluvial transportation, burial and diagenesis, and metamorphism which transforms clays to micas and other non-clay phases. The metamorphic rock may then be brought to the surface to form soil clays, initiating another cycle. Hydrothermal alteration forms clays in crystalline rocks by the interaction of chemically aggressive solutions which turn the old minerals into clays. Hydrothermal and weathering processes are similar in nature but differ in temperature regime.

magma beneath the ocean floor, causes the formation of many clay minerals and the loss of many elements from the basalts. Ores (sulphides) are deposited at the ocean–sediment interface when hydrothermal solutions reach cold ocean waters and precipitate dissolved elements.

Hydrothermal alteration is a process of dissolution and deposition and thus one of transfer of material.

1.2.7 Clay environments and the clay cycle

One can see that the different origins of clays can be related in a cycle where the initiation of clay mineralogy is inevitably linked to another initiation via a sequence of geological processes. Most clay materials initiate at the earth's surface, either in contact with the air or with covering water bodies. Clays form for the most part at interfaces of solid and liquids. They can be transformed by thermal influences until they are no longer clays and enter into the realm of metamorphic rocks. This is largely a question of grain size. The clay phyllosilicate structure is maintained well into the higher-temperature regimes of the earth's upper crust. The interrelated geologic events can be called a clay cycle. Figure 1.4 shows the clay cycle in a diagrammatic fashion.

The clay cycle is renewed when metamorphic rocks are brought to the surface where the weathering hydrolysis interaction creates new clay phases.

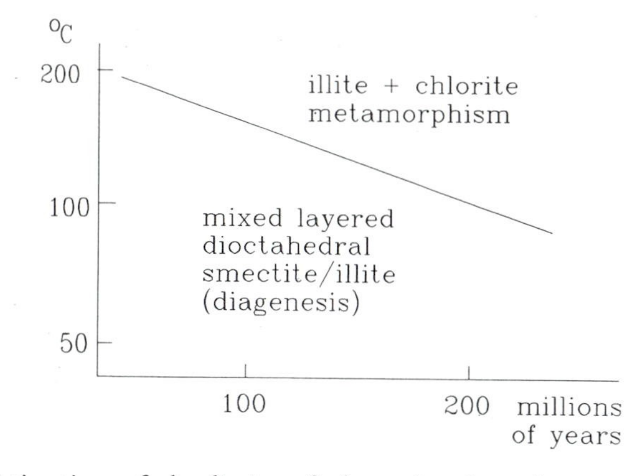

Fig. 1.5 Estimation of the limits of clay mineralogy in terms of time and temperature. For short heating periods clays can be found up to 200°C. This thermal limit is decreased as the heating period is prolonged. Illite-chlorite signals the end of clay facies.

1.2.8 Kinetics

The boundaries of each of the portions of the clay cycle are difficult to fix in the absolute because the processes which change clay minerals occur at low temperatures. Low-temperature reactions are by the nature of kinetic processes slow. Thus, a reaction which begins at, say, 30°C could take tens or hundreds of millions of years to complete. This being the case, the boundaries of clay mineral stability will depend upon the temperature which is applied to the clays and to the time-span over which it is applied. The boundaries will be fixed only when one considers time and temperature. Pressure appears to be a variable of less importance.

We can take the upper boundary of clay mineral stability as an example. The clay–metamorphic limit can be roughly defined as being that of the appearance of the illite–chlorite assemblage which is near to the muscovite–chlorite facies of metamorphic petrology. Figure 1.5 shows the limit of this boundary as a function of time and temperature. It is obvious that the limit between clay stabilities in Palaeozoic rocks will be different from that in Tertiary rocks, something of the order of 120°C or so.

If we look at clay formation at the surface of the earth (weathering), we can see that the concept of time is also important here. Different weathering profiles show that the importance of time is great in the formation of a fully developed clay sequence. Figure 1.6 shows the dimensions of soil clay development on sandy gravels in recent times. These reactions are probably rather slow, other clay profiles in severe environments or based upon chemically highly unstable rocks can be

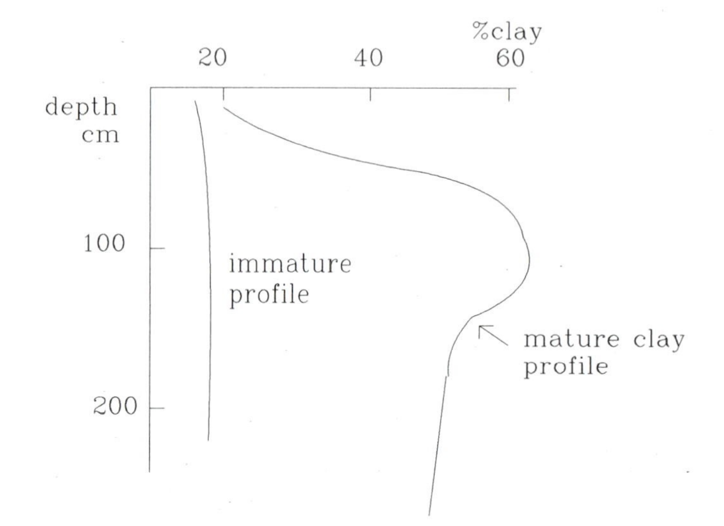

Fig. 1.6 Formation of soil clays as a function of time. The figure shows the increase in clay content going from immature to mature profiles.

developed in hundreds or thousands of years. In general, the older the weathering sequence the more clay is formed. The depth of clay development is also a function of time.

In these two examples it is evident that time is an important factor in the development and extent of clay mineral stability. Thus one must use not only the parameters of chemistry and temperature in describing clays but also time.

1.3 CLAYS AS AN INTERFACE IN THE ENVIRONMENTAL CYCLES

The situation of clay genesis and persistence, low-temperature and surface conditions, indicates that they will be important in what is now called the environment, which is in fact those conditions of pressure, time and temperature which can be associated with human endeavour and survival. Clays are an interface between the surface of the earth, the solid, rock portion of our environment with its aqueous penetration, and that part of the atmosphere which is affected by the activities of man. The problems of pollution via airborne vectors, by surface waste disposal and burial all become involved with the equilibria of clay minerals. The clay environment is the interface between man and his resources (Fig. 1.7). The water which is necessary to his daily activities and growth (industrial, personal consumption and agricultural, plant and animal needs) all passes through

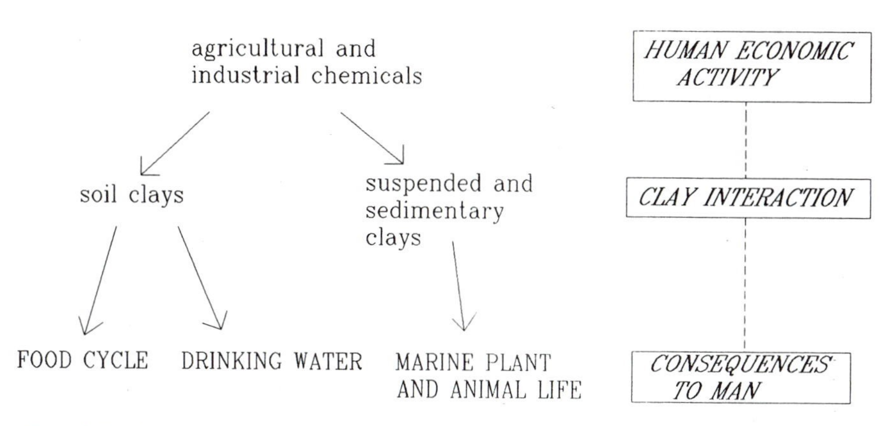

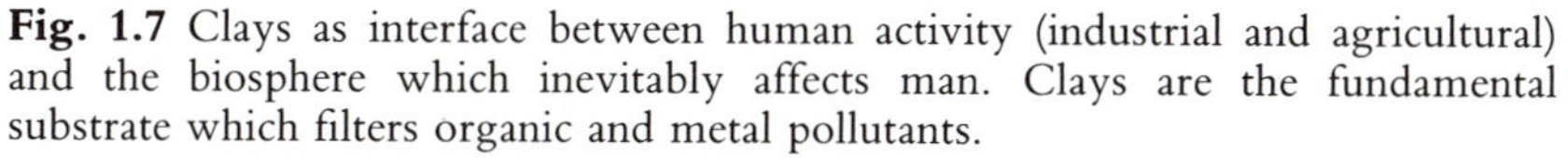

Fig. 1.7 Clays as interface between human activity (industrial and agricultural) and the biosphere which inevitably affects man. Clays are the fundamental substrate which filters organic and metal pollutants.

the clay interface in its cycle of birth and renewal. Rainfall goes through the clay interface, river pollution goes through the clay interface, surface dumping goes through the clay interface and the eventual retrieval of water through wells and river sources represents water which has passed through the clay interface. Thus it is important to understand the chemistry, origin and conditions of stability of clay minerals.

The properties of clays in their interaction with man-made materials will be a determining factor in the changes of these products as they move in a geological cycle towards a sedimentary destiny. The retention rate, the chemical interaction of the clay surfaces with organic and other chemical species will be the key to interpreting their future and longevity in the environment which affects that of modern civilization.

In the chapters which follow we will attempt to give a general outline of the importance of clay minerals: their basic chemical properties and their stability in the environments which will be those of modern man.

2

Tools

Because clay particles cannot be seen with a microscope, the initial petrographic-mineralogical approach, optical petrographic microscopic study, cannot be used. This fact has been a hindrance to the identification, description and systematic study of clay minerals. The foundations of the knowledge of clay structures were laid, and the basic mineral names given, in the late nineteenth and first half of the twentieth centuries. However, much imprecision characterized clay mineralogy. In favourable cases, such as samples with only one mineral type present, it was possible to give a general chemical formula and, with the aid of the initial stages of X-ray diffraction, establish the structural family of clay samples. However, clays are most often found in aggregates containing several different minerals, with millions of grains per gram of sample, and the methods of bulk chemistry or powder diffraction did not permit a sufficiently precise definition of the clay minerals, so that often a mixture of phases was present when one assumed a pure species.

As the normal means of identification could not be used, clay mineralogy had to wait for new technology to come to its aid. The new methods were those of X-ray diffraction, which allows one to determine the phases present rapidly and with precision. Once the phases are known one can proceed to an analysis of the aggregate sample, be it single-phase or multi-phase. Of course, chemical methods are still employed, classical as well as those based upon the phenomenon of X-ray fluorescence in macro or micro (electron microprobe) samples, and infrared spectral analysis and thermoponderal analysis are also very useful in determining and defining the specific characteristics of clays.

2.1 X-RAY DIFFRACTION

The major problem in X-ray diffraction (XRD) is that, since clay particles are very small, a coherent domain in a single crystal will also be very small and will not give strong diffraction maxima, so that most of these maxima will be invisible or only faintly visible. There are insufficient atoms

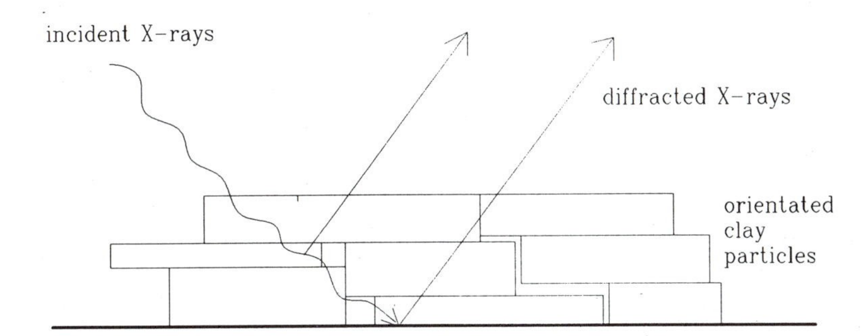

Fig. 2.1 An orientated preparation of clay minerals for X-ray diffraction (XRD). Clays are deposited on a slide so that their basal surfaces (a–b crystallographic direction) are parallel on the plate. The individual crystals, piled one on top of the other, act as a composite crystal, enhancing the diffraction phenomenon.

present in any one plane to give a decent intensity of diffracted X-rays. This is aggravated by the fact that the clays are found essentially in a two-dimensional form, due to their sheet structure. In addition, if the greatest diameter of a clay particle is 2 μm, the smallest is smaller by a factor of 10–20 or more. This reduces the diffraction power of the atomic planes even more in most of the crystallographic directions. As a result, X-ray diffraction had to wait for a clever mechanism to be devised to reinforce the signal coming from the crystal lattice.

2.1.1 Orientated clay specimens

This clever mechanism was the uniaxial powder diffractometer and the use of an orientated clay sample (see Fig. 2.1). The machine uses a parallel X-ray beam of large cross-section which impinges on a thinly dispersed powdered sample (several tens or hundreds of milligrams are necessary) lying essentially in a single plane. The cones of X-ray powder diffraction are intersected by a detector which distinguishes the profile of each diffraction cone on a single circular arc, thus giving a series of diffracted X-ray intensity bands on a continuous trajectory. This is essentially a Debye–Scherrer strip film transformed to a high-intensity trace on paper or now a computer diskette.

Of course, the X-ray diffractometer was not enough in itself. The mechanism of creating a more intense diffraction pattern than would be normal for small crystallites had to be more than a new machine. The trick was to exploit the inherent sheet structure of clays. In orientating the clays, making them lie one on top of the other in the same plane, the diffraction effect is enhanced for the small crystallites. The orientation produces a pseudo-macrocrystal which diffracts approximately as if thousands of the crystallites became a single large crystal. The resulting diffraction characteristics are thus enhanced in the crystallographic direction parallel to the sheet structure. However, the drawback is that the other diffracting planes in the crystals which are not parallel to the sheet structure are almost totally lost to the diffraction spectrum. The X-ray beam sees only the planes parallel to the direction of orientation.

As a result, the identification of clay minerals is done almost entirely on the spacing of the crystallographic planes parallel to the sheet structure. This is called the **basal spacing** of the clays and is of fundamental importance to any discussion of clay mineralogy.

A second happy coincidence allows one to use the orientation properties of the clays to great advantage. This is the characteristic of some clay types to change their cell dimensions in the direction of the sheet structure by swelling. The incorporation of different polar molecules changes the basal spacing of expanding clays, which allows a more precise definition of these clay types. X-ray identification then often involves a manipulation of the crystal composition, by inserting or removing different molecules from between the loosely held sheet layers. Therefore, a complete identification of a clay aggregate necessitates several X-ray diffraction identifications. These operations determine the swelling properties of the clays.

The two standard polar molecules currently used to effect swelling are H_2O and ethylene glycol, which give a 5 Å and 7 Å increase in sheet thickness, respectively. The three states which are characteristic of a clay are a completely **dried sample** (heating above 200–300°C is necessary), **glycollated** and **water-vapour-saturated**. Differential retention rates of the water and glycol molecules by the various clay species gives one an additional means of identification of the clay species present.

2.1.2 Preparation of clays for XRD

Preparation of an orientated clay sample can be done in several ways. The first step is to disperse the sample in water. It is common practice to subject the sample to ultrasonic vibration to dissociate the clay particles in the solution. Grinding and vigorous shaking of the water–clay mixture are more classical methods. Size separation of the clay fraction can be done by letting the dispersed sample stand in a recipient greater than 10 cm deep overnight. The material left in the water will be mostly less than 2 μm in diameter. Finer or more precise size fractioning must be done by centrifugation. The suspension should then be deposited gravimetrically on a flat surface to be presented to the X-ray diffractometer. There are various ways of doing this.

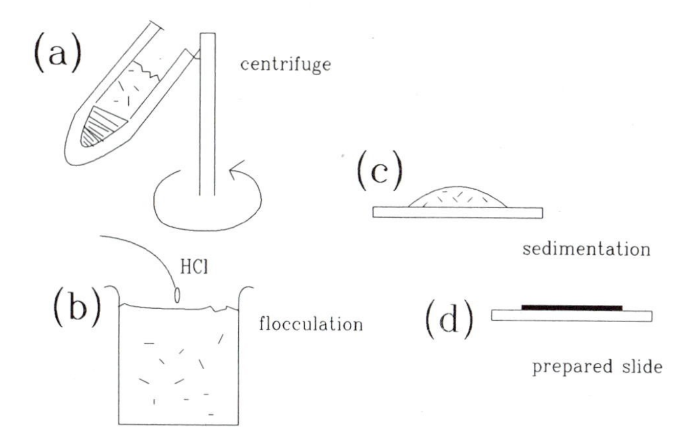

Fig. 2.2 The different steps in preparing an orientated clay sample by settling from a concentrated suspension. Concentration of an initially dilute clay suspension is accomplished by (a) centrifugation or (b) flocculation. (c) Sedimentation of the concentrated clay suspension on a glass slide. (d) The final, orientated preparation after drying.

The suspension can be concentrated by continued centrifugation until the material is deposited at the bottom of the recipient (see Fig. 2.2). This concentrate is resuspended in a smaller volume of water, poured on to a glass slide and allowed to dry slowly to produce an orientated deposit. If time is short or a powerful centrifuge is unavailable, the process can be speeded up by adding a very small amount of electrolyte to the suspension and heating it for several hours. This operation flocculates or combines many clay particles into an aggregate, which can then settle and be concentrated by pouring off the clay-free water. The concentrate is then sedimented on the glass slide. The orientation effect in this method is less efficient than in the case of non-flocculated preparations but quite adequate for most needs.

A second method is to deposit the full suspension on a porous porcelain plate (of dimensions similar to the glass slide) by centrifugation of the suspension through the plate in a centrifuge (see Fig. 2.3). The plate–clay preparation is X-rayed. This method is probably the most reliable but it is also the most cumbersome.

A third, and most efficient, method is to draw the entire suspension through a filter (mylar or other plastic substrates perforated with calibrated holes) in a vacuum (see Fig. 2.4). The deposit is transferred on to a glass slide when wet and, upon drying, the transfer is effected.

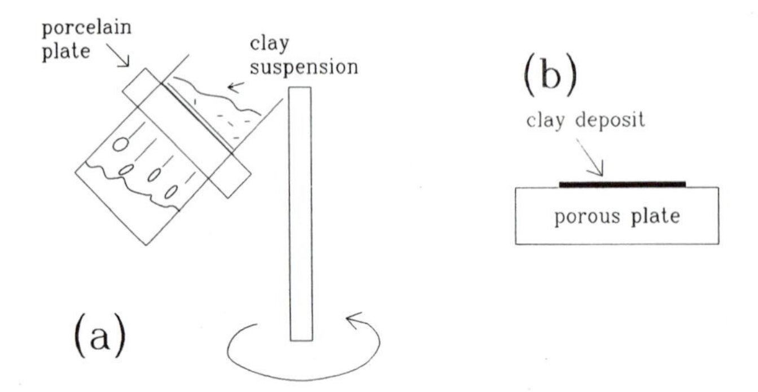

Fig. 2.3 Porous plate technique with (a) deposition of the clay from a suspension drawn through a porous plate by centrifugation; (b) the orientated sample.

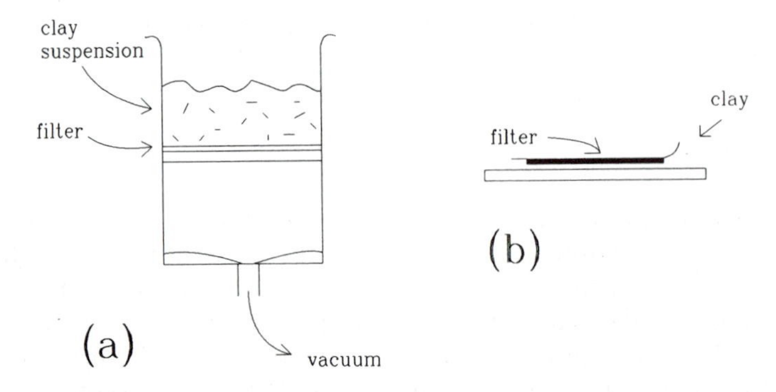

Fig. 2.4 Filter method of preparing an orientated clay sample. (a) Clay suspension being drawn through a filter by vacuum suction. (b) Transfer of the clay from the filter to the glass slide used in the diffractometer.

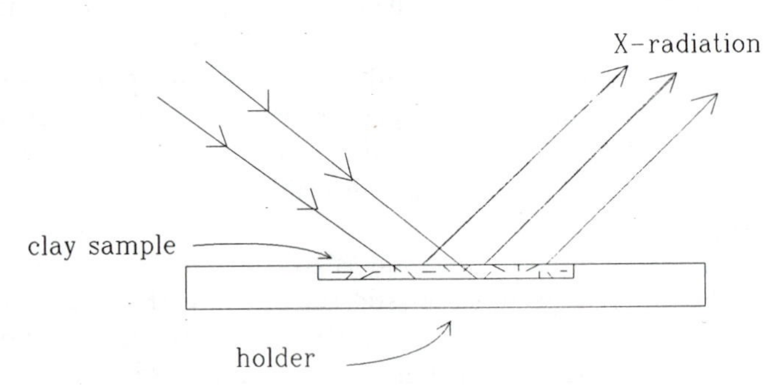

Fig. 2.5 Non-orientated clay sample deposited in a cavity on the surface of a slide.

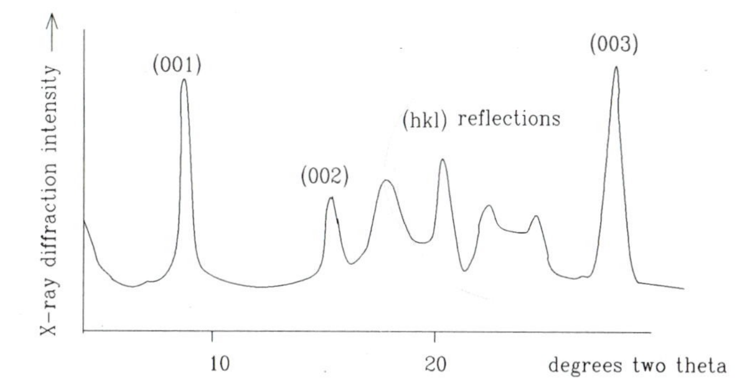

Fig. 2.6 An XRD trace showing the relative position of the basal spacings (001, 002, 003) and the (hkl) bands typical in clays. The X-radiation used is from a Cu anode tube.

2.1.3 Non-orientated clay specimens

Disorientated or non-orientated preparations are necessary to determine diffraction maxima other than the (001) basal spacing series. Since the clays have a natural tendency to be orientated, they most often give poor (hkl), or non-basal, reflections. The sample must be ground finely and lightly packed into a shallow cavity in a metal plate. Care must be taken to avoid pressure orientation of the powder. There are other preparation methods, but everything depends upon the care taken to avoid any physical pressure on the particles (see Fig. 2.5).

2.1.4 XRD spectra

In the XRD spectra of clays there are several diagnostic zones which give characteristic diffraction maxima. The powder spectra are usually obtained using an X-ray tube with a copper anticathode. These have been the most reliable tubes over the last three decades. However, other tubes can be used to great effect, such as those with a cobalt anticathode. Figure 2.6 shows the major zones of interest in a typical spectrum of an unorientated or partially orientated sample of clay. The first two diffraction maxima are the first- and second-order basal reflections, (001) and (002). These are followed by a zone of (hkl) reflections which is more or less complex depending upon the degree of order in the stacking of the basic layers of the structure in the clay crystallites. The more peaks in these zones the more complex the stacking arrangement of the sheet units. The stacking sequences are given names as specific structural **polytypes**. The (hkl)-rich zones occur on either side of the third-order basal reflection.

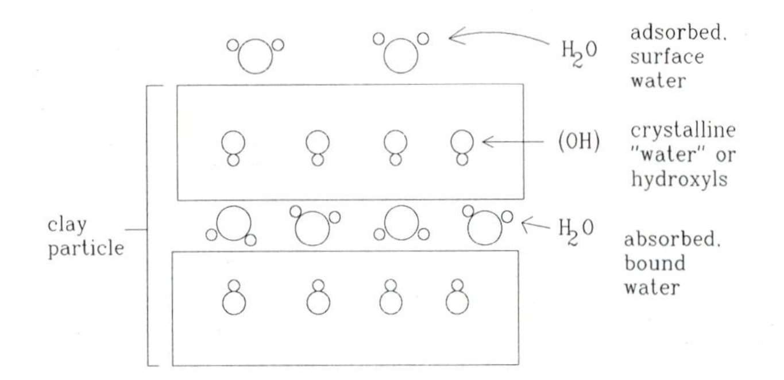

Fig. 2.7 Illustration of the types of adsorbed water on the surface of the clay crystallites: bound absorbed water found between the clay layers and crystalline water (OH units) found within the clay layers.

Another zone, not on the figure, is that where the (060) reflection can be identified, around 1.5 Å or 60° 2θ using copper radiation. This reflection is diagnostic for several structural and compositional determinations.

Most XRD diagrams begin at about 2° 2θ (30 Å) and run to about 35° 2θ (about 2 Å) using copper radiation. Usually a short interval around 60° 2θ is run also to determine the (060) bands.

2.2 THERMO-GRAVIMETRIC ANALYSIS

Heating of clays at a regular rate effects weight loss depending upon the retention of water in or on the clay structures. All clays have at least two, at times three or four, types of water or hydrogen ion in or on their crystallites. The binding energy of these types of water can be used to identify the clay minerals. One analysis method is called thermo-gravimetric analysis (TGA). The different types of water of hydrogen atoms bound to the clay are shown in Fig. 2.7. Several tens to hundreds of milligrams of sample are necessary.

The first type of water lost is **adsorbed water**. This form, which is held by less strong attractions, is found on the surface of the clays in defect sites or at sites of broken bonds of the silicate structure. It is usually present in small quantities which are proportional to the surface area of the clays. Since all clays have roughly the same outer crystalline surface area, the amounts of adsorbed water are similar from one species to another. There is little diagnostic value in the determination of this quantity. Most clays are analysed after this surface water has been eliminated by heating to 80–90°C. In a DTA diagram (see Section 2.3) of an untreated sample, the initial weight loss of 1% or so is due to this water.

The next type of water is of the **zeolite** type. Unlike adsorbed water,

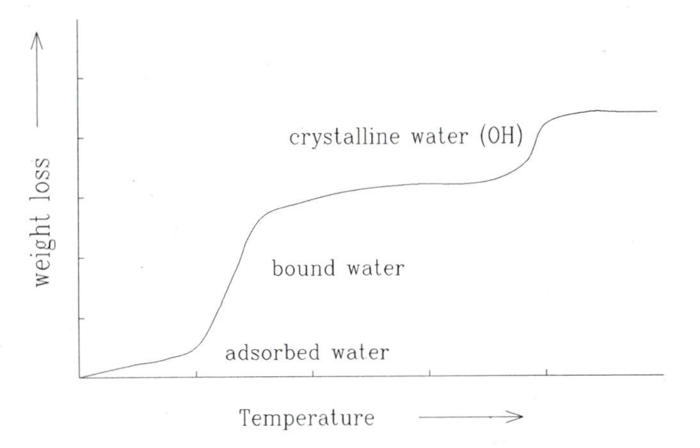

Fig. 2.8 Illustration of the weight loss–temperature curves found for smectite and palygorskite-sepiolite clays. On the curve one can see the small weight loss of the adsorbed water at low temperatures, important weight loss of the absorbed water and a small weight loss of the crystalline water. Each type of water loss at higher temperatures indicates an increasing bonding energy of the water on to or in the clay surface.

the zeolite type is found within the structure in approximately fixed quantities. Such water is found in only one clay mineral group, the sepiolite-palygorskites, and it does not appear in Fig. 2.7. This water leaves the structure at 100–150°C.

Bound water, associated in a geometric structure around a cation, is found between the sheet layers of smectites. This water is structured by its coordination with the cation in a one (2.5 Å thick) or two (5 Å) water-layer unit which is found to lie between the sheets of smectites. This water normally leaves the structure at 100–200°C. Sepiolite-palygorskites also have bound water in their structures but its loss does not change the cell dimensions of the clay.

Crystalline water is found within the sheet layers, as (OH) units. This water is more firmly bound to the structure, and temperatures of 500°C or more are necessary to free it. A typical TGA diagram is shown in Fig. 2.8 with the three major types of water – adsorbed, bound and crystalline.

2.3 DIFFERENTIAL THERMAL ANALYSIS

This method of analysis is based upon whether the thermal reaction which occurs as a clay mineral is heated is exothermic or endothermic. Here not only water loss (endothermic reactions for the most part) but also eventual recrystallization and recombinations (exothermic) are observed (see Fig.

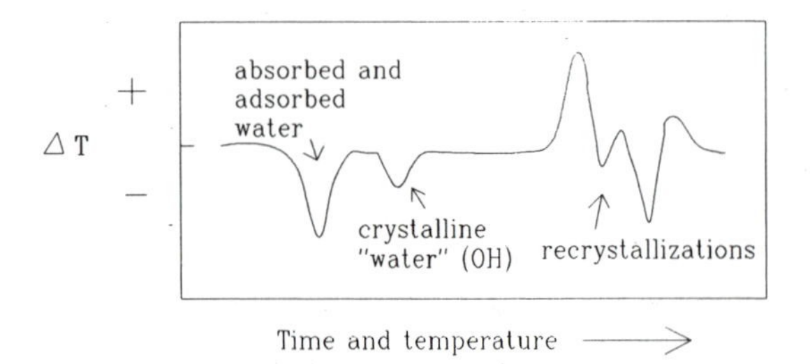

Fig. 2.9 Differential thermal analysis (DTA) diagram showing the exo- (+) and endothermic (−) reactions of water loss and recrystallization occurring in clays as temperature and duration increase.

2.9). The complexity of the full mineral transformations and the problems of kinetics concerning these reactions makes interpretation of full DTA diagrams very complex. Such methods are rarely employed in routine investigations; they are left for the more special applications of mineral species characterization or certain industrial applications such as ceramics.

2.4 INFRARED SPECTRAL ANALYSIS

Infrared (IR) spectral analysis has been used to characterize clays for some time, and to a lesser extent to identify them. This method is somewhat less sensitive to the differences in clay structures and as a result its use as a method of identification, especially in a mixed-phase sample, is difficult. However, the method is useful in identifying certain features of pure phases. An XRD study is usually necessary before this method is used. In IR studies, less than 1 mg is used to obtain a rapid spectrum and as little as 0.1 mg can routinely be used.

In clay studies IR spectra are produced by passing a multi-wavelength infrared beam through a finely dispersed sample. The normal method uses a clay dispersion of less than 1 mg in several hundred milligrams of KBr. The mixture is subjected to very high pressure to vitrify the KBr and leave only the clay as a crystalline material. Such dispersions are stable to 600°C and thus heat treatment is possible to observe the various states of water or hydrogen ion bonding in the clay.

The clay structure absorbs the IR radiation according to the vibration frequencies of its various crystalline components. These range from the OH unit through the SiO_4 and AlO_4 tetrahedral units, the AlO_6, MgO_6, FeO_6, etc., units to complex, multi-atomic portions of the silica sheet network.

The smaller the number of atoms in the vibrating unit, or the lower the mass number of an atom, the higher the energy of vibration and the

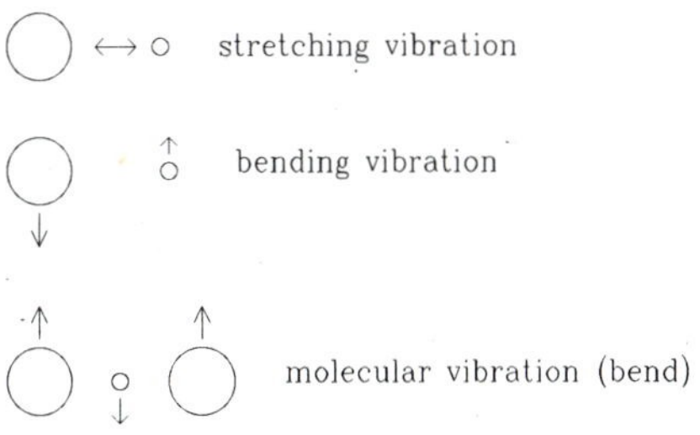

Fig. 2.10 Ionic vibrations of atoms detected in crystals using infrared (IR) spectral methods. Vibrational modes are OH stretching; Si, Al, Fe tetrahedral stretching (Si tet), Mg octahedral movements (Mg oct) and bending motions of silica tetrahedral ions (Si-O-Si).

shorter the wavelength absorbed by the vibrator. By convention, IR spectroscopists use the inverse of the wavelength in centimetres (cm^{-1}) to designate the frequency of vibration.

Figure 2.10 shows the vibrational modes which are used to describe the atomic vibrations. The stretch and bend modes are the most important in that they involve a small number of atoms and thus can be used to identify the components and be useful in specifying the occupancy of certain sites in clay structures. The stretch mode is the most direct and the highest-energy vibrational mode. It involves the motion of one atom against another or of several atoms against a single atom. The bending mode is less energetic, involving oblique movement directions of the atoms with respect to the positions of the others in the vibrating unit. Figure 2.11 shows a typical IR spectrum for a clay mineral.

The most useful zones of IR observational spectra for clay minerals are in the OH vibrational region because here one can have access to information not available from other analysis methods.

It is of little use to find the chemical composition by IR spectra when there are other, easier methods available which are more precise – the electron microprobe, for example. However, the bonding energy of the crystalline water (OH units) in the clay structures can be determined with ease using IR methods. This information is difficult to obtain by any other process. Using this method the sample is heated to over 400°C and kept dry to avoid adsorbed and absorbed H_2O.

The vibrational energy of the OH units, in fact the hydrogen which vibrates against its linked oxygen, is a function of the immediate cation neighbours of the OH site. The most important are the octahedral ions, which are found in arrays of two or three ions (Fig. 2.12) around the OH groups in the octahedral site (see definition of this site in Fig. 3.4). The interaction between the octahedral layer cation and the oxygen anion

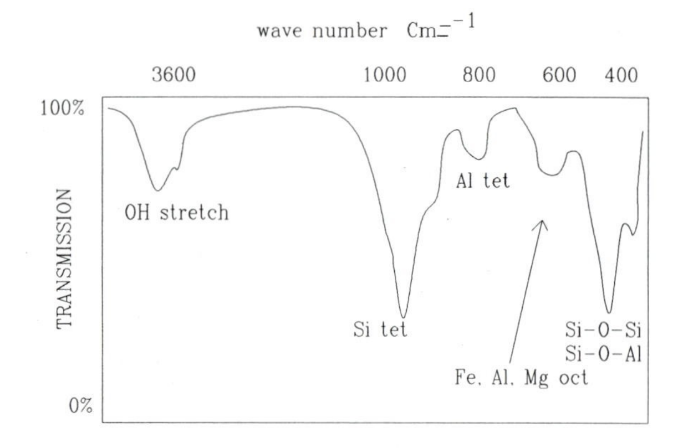

Fig. 2.11 Typical infrared diagram showing the different frequency regions of the characteristic vibration types shown in Fig. 2.10. OH strength vibrations have the highest energy, followed by Si-O and Al-O tet vibrations, Fe, Al and Mg oct vibrations and complex Si-O-Si tet vibrations.

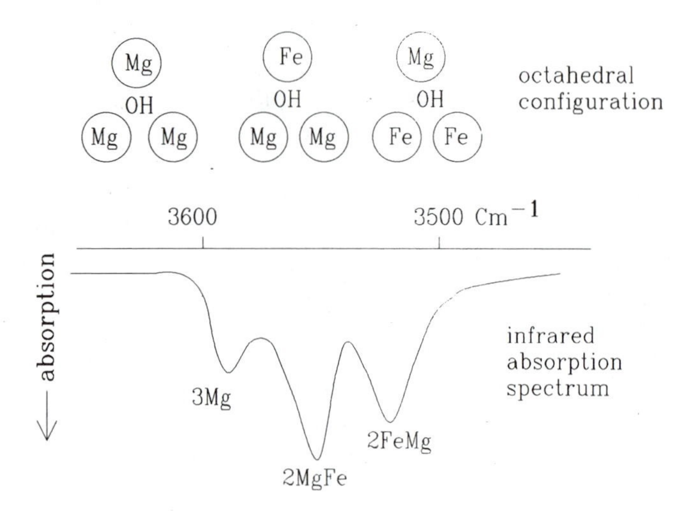

Fig. 2.12 Illustration of some possible octahedral ion configurations which lead to characteristic OH vibrational frequencies. The iron atoms produce bands with lower wave numbers than do the magnesium atoms.

determine the strength of the oxygen–hydrogen bond. IR spectra show the bonding energy of the OH bond in the sheet silicate structures. Therefore IR spectra give an indirect, but accurate, determination of the octahedral site occupancy.

The adsorbed water in clays can also be investigated using IR methods, but with more difficulty. It is difficult to distinguish between inter-layer, cation-coordinated adsorbed water on the clays and the water adsorbed on KBr. Therefore the determination of the structural state of the inter-layer, cation-coordinated water is best investigated using other substrates for the clay sample. These methods are not routine methods of investigation, as one can imagine.

One should remember that IR spectral analysis is particularly useful in determining the environment of the crystalline water (OH) in clays and therefore the arrangement of the octahedrally coordinated cations.

2.5 ELECTRON MICROSCOPES

These methods of analysis are based upon the physics of interaction of a beam of electrons with crystalline (or non-crystalline) matter. Several derivative methods of analysis are based upon the same principle. The source of probing energy is a beam of electrons generated much as light is generated in a lightbulb. A thin wire of a resistant material is linked to stream of electrical current. The resistance of the material causes it to transform the electrical energy into other forms (light in the case of a lightbulb). However, in our new-generation machines, the energy sought is the excess of electrons bottled up in the thin resistant wire. These electrons are expelled from the circuit and attracted elsewhere in the apparatus by high electrical potentials. The escaped electrons are shaped into a finely focused beam by electromagnets. This beam is directed on to a target of interest (clays, for example) where different physical processes of interaction take place and, depending upon the apparatus, analysed by different methods.

2.5.1 Scanning electron microscope

The scanning electron microscope (SEM) uses the flux of secondary and backscattered electrons from the material to form an intensity image of the material bombarded by the electron beam. The resulting cathode tube realization indicates the three-dimensional aspect of the sample. This is very useful in the identification of textures and shapes of mineral grain aggregates. The definition or resolution of the image is of the order of 0.01 μm.

Much work has been done with SEMs in the characterization of clay textures in sedimentary rocks in an attempt to identify the physical properties which the clays create in sedimentary rocks. For example, it is very important for a petroleum engineer to know why oil does not flow from a rock which has an apparently large capacity for liquids. In some cases the pore spaces are filled with clay particles which do not take up much room but which inhibit the movement of hydrocarbons.

Figure 2.13(a) shows different clay mineral shapes which can be seen using an SEM. The sample was formed at high temperature in the laboratory. The photo shows blades of talc (lower part of the photo) with a cap of filamentous serpentine called chrysotile (top of photo). Figure 2.13(b) shows a detail of the talc crystals while 2.13(c) shows a close-up of the chrysotile 'threads'. The second series of photos (2.13(d) to 2.13(g)) shows the morphology of clays as they are presented under the conditions of the microscope (that is, under high vacuum). Typical features such as kaolinite 'books', fibrous I/S minerals and tightly intergrown chlorite can be seen growing on quartz grains in the illustrations. The SEM shows clay crystal morphologies very clearly.

2.5.2 Transmission electron microscopes

Transmission electron microscopes (TEM) are used to see the shapes of crystals in essentially a two-dimensional plane. The clays are deposited on a carbon-coated copper grid (with holes in it so that the clays can be 'seen' by the electron beam). The electron beam is absorbed more by the clay than by the carbon film and thus a shadow image of the clay particle is

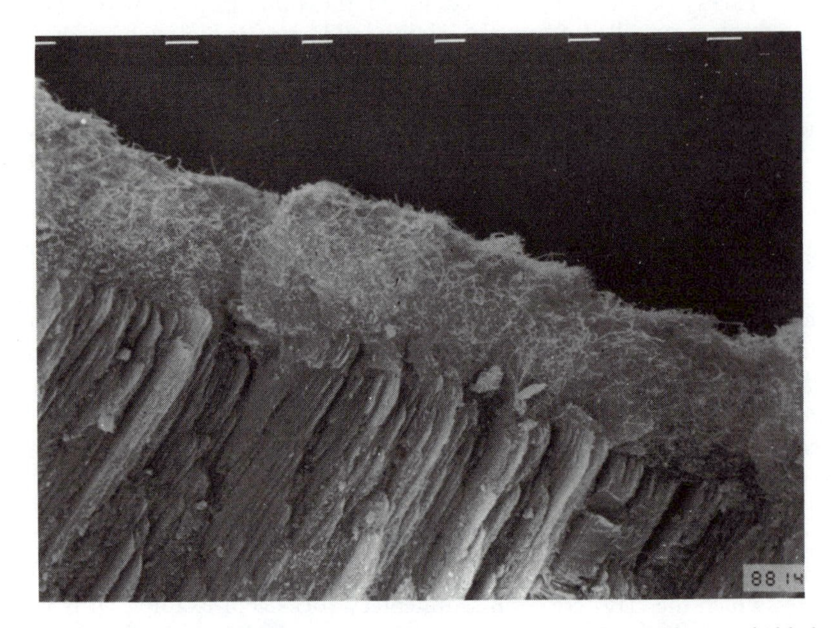

Fig. 2.13 SEM photographs of some synthetic clay minerals. (a) Shows talc blades with a surface of chrysotile overgrowth.

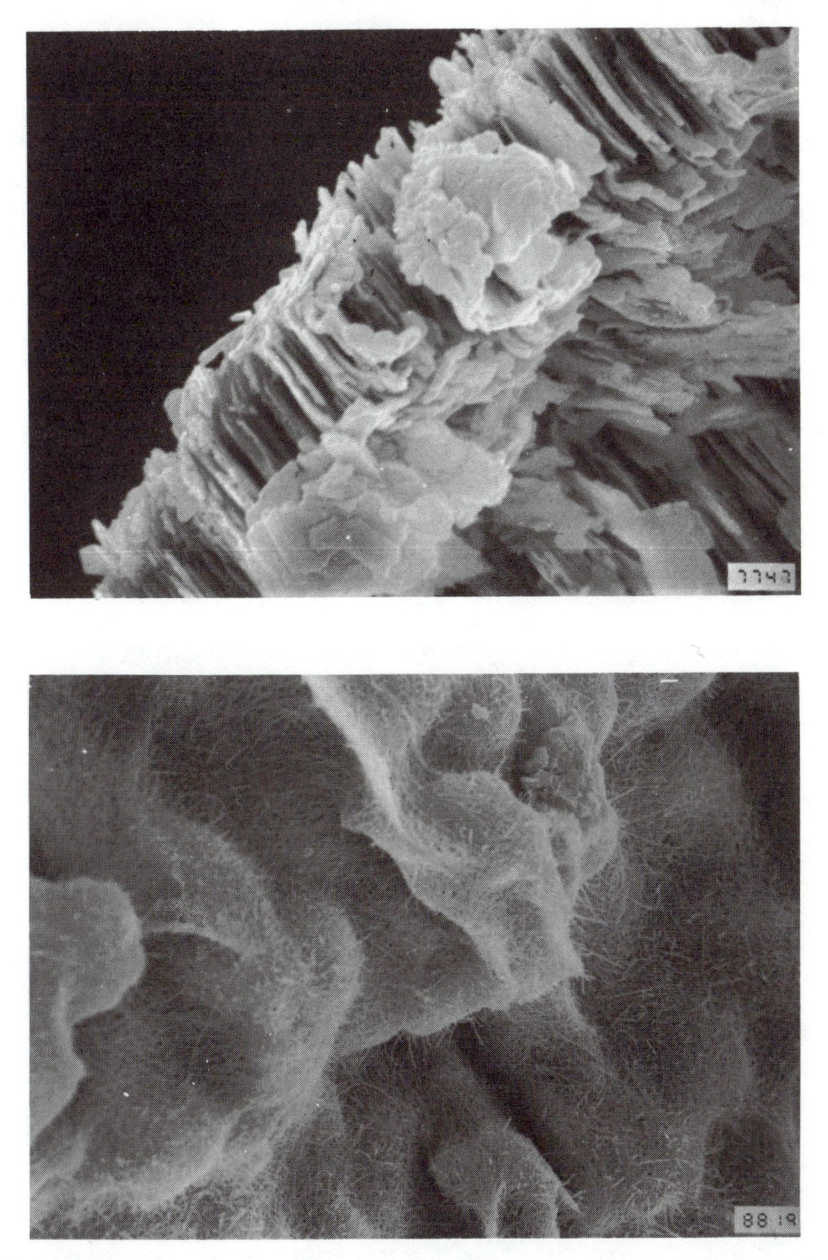

Fig. 2.13 (cont.) The flat, sheet structure of talc is seen in (b) where the blade-shaped crystals are evident, while the needle-like chrysotile structure is well demonstrated in (c).

Fig. 2.13 (cont.) (d) Shows the thin, blade or lath shape typical of the ordered mixed-layer illite/smectite mineral. (e) Shows a dried aggregate of these crystals. One can see the lath-shaped crystals on the edges of the aggregates.

Fig. 2.13 (cont.) In photo (f) the book-shaped habit of kaolinite is evident. Hexagonal sheets are well expressed in the crystal morphology. Photo (g) shows aggregates of illite/smectite crystals in the foreground which are found as an overgrowth phase on stubby blades of chlorite found in the background of the photo. The illite/smectite is a later phase than the chlorite in the crystallization history of the diagenetic minerals of the sandstone in which they are found.

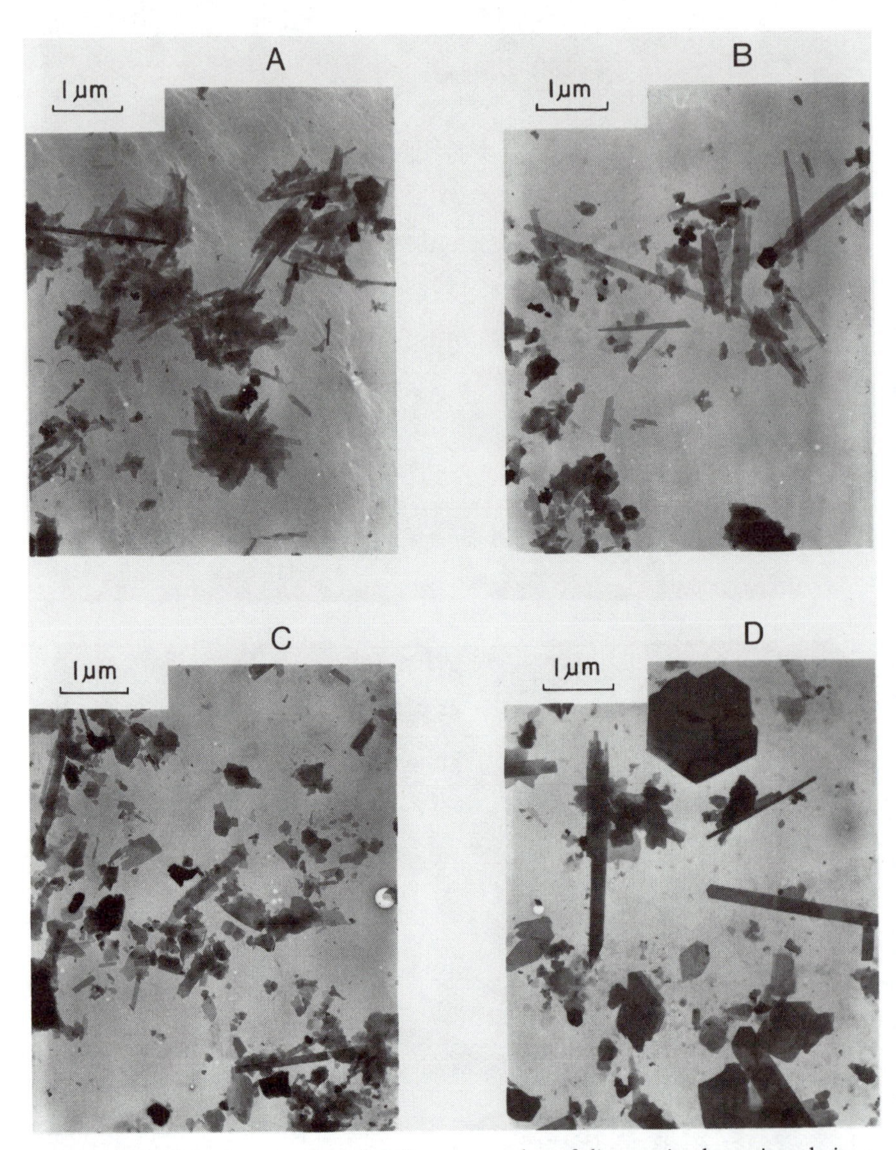

Fig. 2.14 TEM photographs of different samples of diagenetic clay minerals in a growth sequence. In photos (a) to (d) one sees that the crystals become larger and better formed as growth proceeds at greater depths in a sedimentary rock sequence. The thin blade-like crystals are illite/smectite minerals and the hexagons are illite crystals. The illite grains become larger and more abundant at depth as growth proceeds. The illite/smectite laths grow also but become relatively less abundant, indicating that some of them dissolve to form the illite hexagons and 'help them to grow'.

produced. The TEM resolution is greater than that of SEM microscopes (μm). Their main use is that of identifying the crystal shapes of clays and their dimension. Also, the detection of fluorescent radiation by energy-dispersive techniques (as on electron microprobes) can be used to estimate the composition of individual crystallites. Figure 2.14 shows some crystal shapes seen in clay minerals. Most clay structures have a hexagonal symmetry which gives either hexagonal forms or those derived from a hexagonal prism. Some crystals are sufficiently elongated to appear as needles.

2.5.3 High-resolution transmission electron microscopes

The newer electron microscopes give a new smaller dimension to the study of clays. Transmission micrographic methods give resolutions of less than 2 Å. This allows one to 'see' the levels of atomic planes in the clays with ease and in special cases the atoms themselves. Using energy dispersion analysis methods it is relatively easy to determine the composition of a clay mineral when it has a dimension of several unit layers, that is, greater than 100 Å. It is also possible to determine the homogeneity of an individual mineral crystal on a scale previously undreamed of. The succession of unit layers, the intimate interlayering of the clay structures and their growth patterns, are now easily determined using the high-resolution electron microscope (HRTEM). Figure 2.15 shows several of the features which are accessible using such equipment.

The drawback of HRTEM studies is found in the fragile nature of the clay minerals which absorb too much of the electron beam and convert it to thermal energy. The time during which one can take a 'picture' of the clay mineral atomic layers is very short and many poor photos are made. Several dozens or hundreds of photographs may need to be taken to obtain one which has the focus and sharpness needed to convince the clay mineralogist of structural details. HRTEM is not a routine method of clay analysis but it is a very crucial one in determining the structural relations of a well-defined clay material. Figure 2.16 shows the definition one can expect in HRTEM studies of clays.

2.6 ELECTRON MICROPROBE

The electron microprobe is essentially an X-ray fluorescence spectrometer hooked up to an electron microscope. The idea is simple enough but the physics of the electron–solid interaction and the ensuing escape of the generated X-ray fluorescent radiation have taken quite some time to master, at least thus far. The method is a reliable and reasonably rapid one which can be used to obtain an excellent chemical analysis of a very small quantity of clay material. The diameter of the analysis spot is near 2 μm

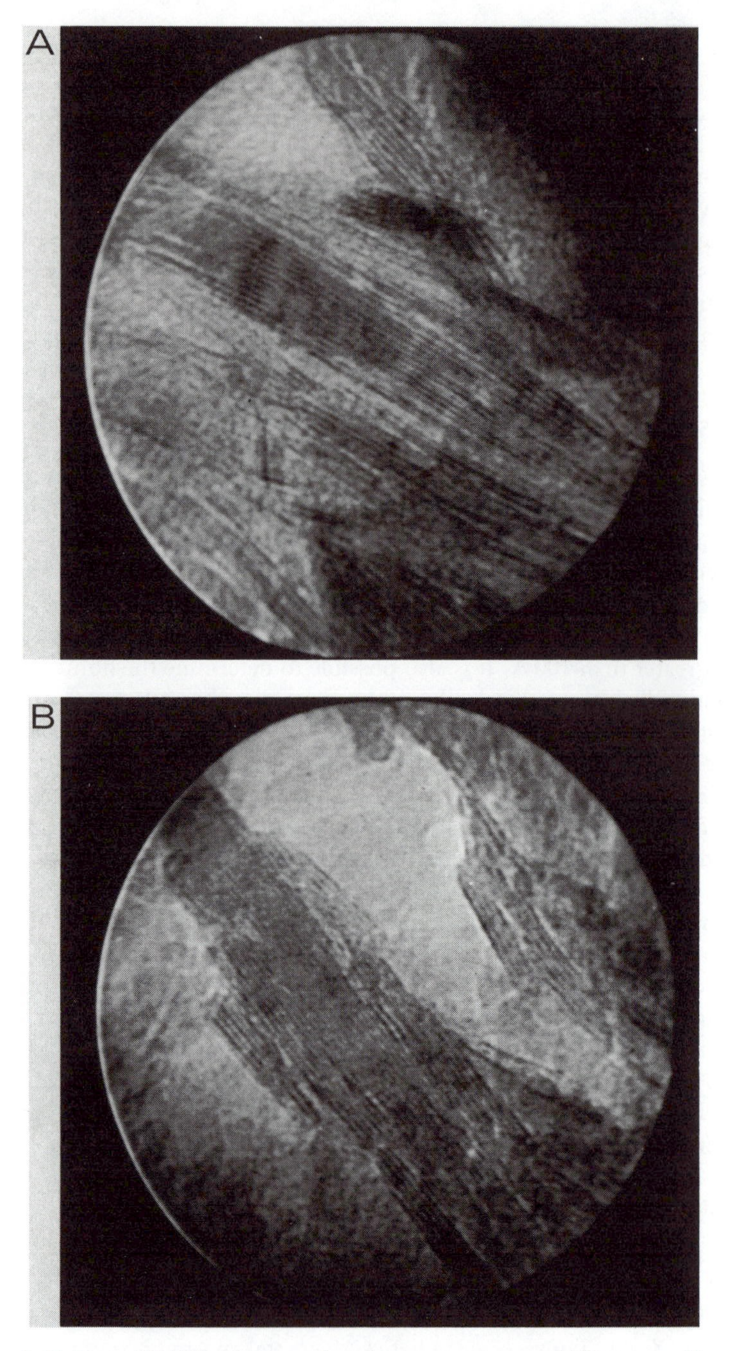

Fig. 2.15 Two HRTEM images showing two aspects of illite crystallites. The sample is prepared perpendicular to the sheets of the crystals and the resultant image shows the 10 Å layers of the illite structure as bands within the crystal. In photo (a) the crystals form separate grains. In (b) the particles are multigrain, being composed of several small crystallites of five to ten layers.

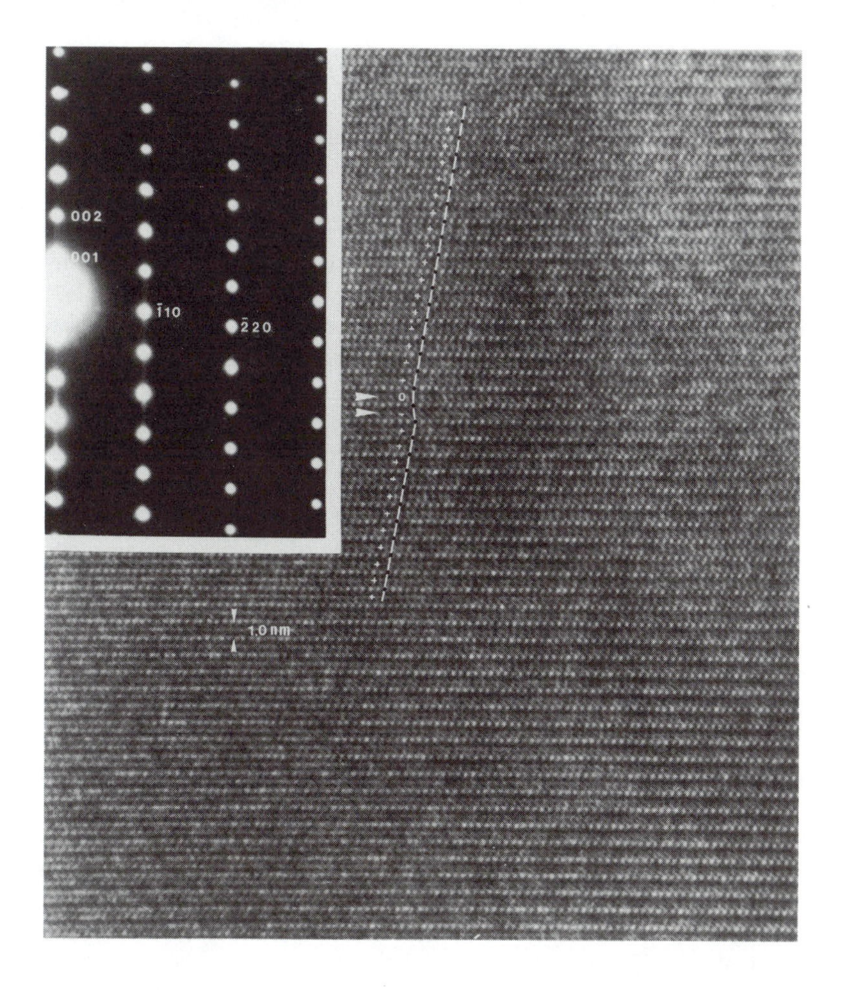

Fig. 2.16 Biotite crystal viewed down the (110) crystallographic direction. A single stacking (growth) fault is seen in the middle of the line showing the crystallographic continuity. A micro-diffraction image is shown giving the diffraction spots representing the crystallographic planes (inset upper left). The small white circles in the photo are the silica tetrahedra and the small ovals the octahedral coordination configurations. In this picture one approaches the atomic scale, but not quite. One can see the great atomic regularity of the biotite structure. In true clays the structure is not as large and photographs of this type are much more rare. There is not enough atomic coherence to permit such a photographic image of clays (photo courtesy of A. Baronnet, University of Marseilles).

and the penetration of the beam which excites detectable radiation is about 6 μm.

This method of analysis allows one to use a normal petrographic thin section (thickness 30 μm) to locate the precise environment of clay formation or the precise grains which give rise to new clay minerals. One can have a spatial context of clay formation. In weathering, for example, it is possible to see what clays form from which minerals. Results are listed in the normal chemical fashion, giving oxide percentages or atomic proportions. The method is now routine. The resolution of the method is close to that of the petrographic microscope and therefore the old limitations are still present, that is to say, a single clay particle cannot be designated for analysis as in TEM.

Two main types of detection of the electron-excited radiation (X-ray fluorescence) are commonly used, wavelength dispersion and energy dispersion. The first system extracts a portion of the generated secondary X-radiation (X-ray fluorescence) into a crystal spectrometer. The system uses the Bragg principle 'in reverse' compared to XRD analysis. The white (multichromatic) radiation generated is focused onto a crystal which diffracts it (as a prism would diffract white light) so that the wavelengths of the characteristic radiations due to the elements in the sample are dispersed in space. A sensitive detector is displaced around the diffracting crystal so that the intensity of the different wavelength radiations is recorded by the machine. Using appropriate corrections, the raw intensities are converted into the initial excited intensities due to the atomic ratios in the target.

Energy dispersion uses the energy of the radiation photons as their method of identification, instead of their wavelength. The conversion of energy into an appropriate signal of an appropriate intensity is done in a crystal similar to a common p-n transistor, a lithium-drifted silicon crystalline material. The p-n gap determines the energy range at which the detector is most efficient. These detectors are more sensitive (efficient) than the wavelength systems but they give a higher background (not element-specific radiation) and hence are less precise than the wavelength methods. For a weak signal, low electron beam current, energy dispersion is more efficient and for a high beam current the wavelength method is better. In the analysis of clays, low beam currents are normally better, to avoid destruction of the loosely bound, hydrated, alkali-bearing structure. Therefore energy dispersion is better in many cases.

Figure 2.17 shows an SEM – energy analysis of chlorite crystals. Above one sees the SEM image of the chlorites, and below, in part (b) one sees the energy spectrum of the elements in the chlorites. The peaks to the left show Mg, Al and Si concentrations and those on the right Fe and the copper substrate of the sample preparation.

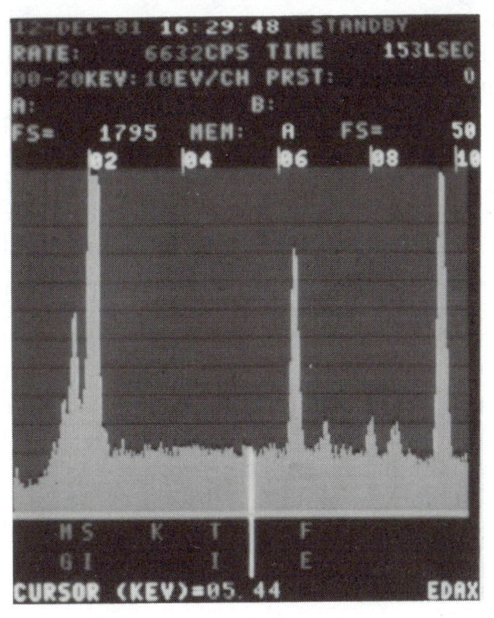

Fig. 2.17 Chlorite blades growing on a quartz grain. In photo (a) one sees a square which delimits the area of analysis where an energy dispersive system of X radiation analysis was made. The spectrum obtained is seen in (b). The relative intensity (height) of the characteristic radiation is related to the relative atomic proportions of the elements present in the chlorite. The chlorite is iron-rich.

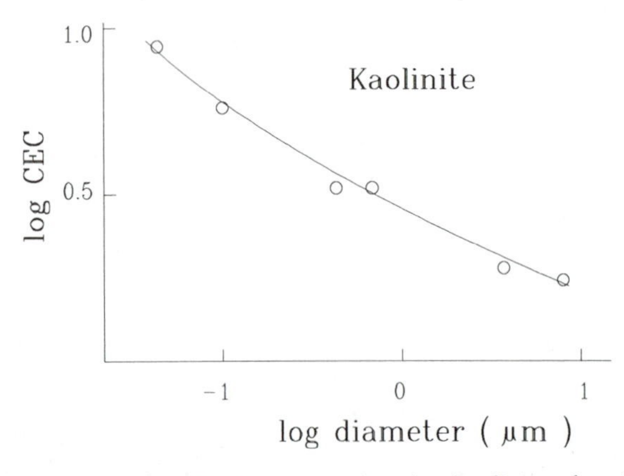

Fig. 2.18 CEC and particle diameter plot for kaolinite showing the relation between increased grain surface and increase of cation exchange capacity (CEC) due to surface bonding.

2.7 CATION EXCHANGE CAPACITY

Cation exchange capacity (CEC) measures two of the fundamental properties of clays, the surface area and the charge on this surface area. The surface of a clay can be of two sorts, external and internal. The external exchange capacity measures nothing more than the average crystalline size. There is little specific chemical information in this measurement. The surface capacity of adsorption is largely dependent upon broken bonds and surface growth defects. Figure 2.18 indicates the importance of such defects as they are created by diminishing the grain size of a clay mineral. CEC increases as grain size decreases.

The internal exchange capacity is much more interesting in that it reflects the overall charge imbalance on the layer structure and the absorption capacity of the clays. The exchange capacity is an estimate of both the number of ions absorbed between the layers of a clay structure and of those adsorbed on the outer surfaces.

Without going into the details of the structure and ionic substitutions in clays, which will follow, it can be said that the exchange capacity is almost always measured as a function of the number of cations (positively charged) which can be measured on the clay surface once it is washed free of exchange salt solution. The operation is performed by immersing a quantity of clay in an aqueous solution containing a salt, usually chloride, or ammonium hydroxide. The clay is extracted from the solution after being washed several times to ensure that there is no salt present and therefore that all of the cations are fixed on the surface having been exchanged for ions already residing on or in the clay (hence *exchange*

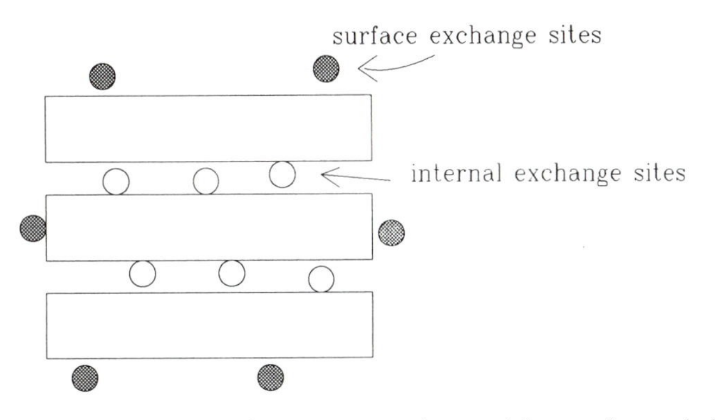

Fig. 2.19 Different types of exchange sites on clay particles, surface and absorbed ion interlayer sites. These exchange sites are quite similar to the sites of adsorbed and absorbed water on swelling clays.

capacity). The operation is done to homogenize the species of ions present in order to facilitate the identification of the amount of material present as exchangeable, labile cations. Usually the exchange ion is one not likely to be present in the sample naturally so that any cation which remains and which is not exchanged can be measured. The favourite ion used by clay mineralogists is strontium.

The quantity of ions present is recorded as the charge, milliequivalents, which are found to be fixed on 100 g of dried (usually fired to 110°C) clay. This is not an easy unit of measure because the charge must be related to the atomic molecular weight in order to have any meaning with relation to a mineral unit formula. However, in mixed-phase samples CEC is a convenient measurement to give an idea of the general chemical properties of a clay-bearing sample. The cation exchange capacity is then given in milliequivalents per 100 g of dried material (meq/100 g or simply meq).

Figure 2.19 shows the positions of exchange ion sites on or within the clay particles.

As it turns out, the measured amount of cation fixed as charge on a clay can vary with the cation exchanged. The usual norm is with Sr as the exchange cation. The charge on the surface of clays in adsorbed sites is usually on the order of 5–10 meq.

Addition of a small concentration of salt to a clay suspension (where particles remain in the aqueous medium through Brownian motion) will tend to form particle aggregates. The process is called **flocculation**. The aggregates tend to become large and dense enough to settle and the clay water suspension is destroyed. This process is independent of the cation exchange by absorption (between the sheets of the clay particles).

The charge measured as being exchanged on sites between the layers of clays or within the structure varies between 40 and 120 meq. This value should be proportional to the charge on the clay layers induced by substitutions of ions which create an electrostatic charge imbalance on the silicate.

One can distinguish two kinds of clay, those with low exchange capacities (10 meq) and those with high capacities (40–120 meq).

2.8 SUMMARY

The tools listed above are those in common use today. Others have been used in the past and certain new methods are beginning to be used in clay mineral studies today. This means, of course, that the panoply of investigative methods will be enlarged in the near future.

Those mentioned above can be classified into the zones and scales of interest in clay structures.

1. XRD and SEM are methods used to study the form and structure of crystal aggregates. Information *can* be derived concerning the crystals by using monomineral samples but *always* by using aggregates of crystals. Average interplanar distances and the average composition of the atomic planes can be derived from XRD studies. The morphology of aggregates is determined by SEM.
2. DTA and TGA studies are concerned with the water molecules in clays. Such studies are done on clay aggregates.
3. TEM investigates the shapes of individual crystals and the compositions of these individual crystals. The observation is looking down on the top of the sheet-like clay grains, that is, the direction of the greatest lateral extension of the crystals.
4. IR methods investigate the relations of individual molecules in the crystals, OH, H_2O, Si-O etc. This is then within the crystals but the method is limited to multi-crystalline aggregates.
5. At the same level or below it are the HRTEM analysis methods. Here the atomic layers or the atoms themselves are visible to the investigator's eye via the electron beam.
6. Cation exchange is used to determine the amount of exchange ions which can be fixed on a clay structure.

The relations of scale and method are summarized in Fig. 2.20.

2.9 REFERENCES

This book is designed to give the basic principles and facts concerning the study and use of clay minerals. Chapter 2 gives the basic methods of

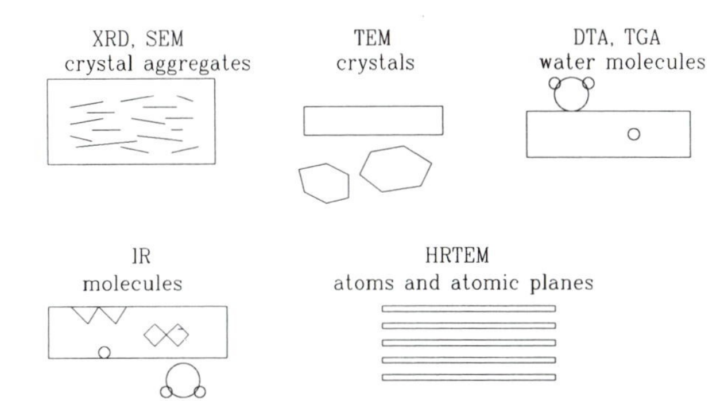

Fig. 2.20 Uses of the different methods of analysis to determine the properties of clays. The sequence indicates that of scale with XRD and SEM (scanning electron microscope) being the largest-scale methods and HRTEM giving the smallest-scale view of clay particles.

investigation currently used in the identification and characterization of clays. One of the most precious tools is the body of information found in books and articles in the professional journals. In any discipline the most important tool is the printed word representing the body of knowledge gathered up to the present time. In the list below one can find the major professional journals which treat clay minerals and clay-related problems as well as several texts treating different aspects of clay mineralogy.

The initial step for any person wishing to gain an insight into a given subject is the use of an abstracting service which treats the body of knowledge published in journals. These abstracting services give the references to published articles by subject matter. For instance, one might look up 'clay minerals', with 'infrared' as a subheading, in order to find information on the infrared spectra of the mineral kaolinite. Another way would be to look up 'kaolinite' and the subheading 'infrared spectra'. There are several abstracting services available, some on computer data banks, and others in printed form. Those below are commonly found in university libraries:

Mineralogical Abstracts, published by the Mineralogical Society of Great Britain
Chemical Abstracts, published by the American Chemical Society.

Professional journals specializing in clays are:

Clays and Clay Minerals, American Clay Minerals Society
Clay Minerals, European Society.

Other journals containing articles on clays are:

American Mineralogist
European Journal of Mineralogy
Contributions to Mineralogy and Petrology
Journal of Sedimentary Petrology
Economic Geology
Geoderma
Soil Science
Journal of Soil Science
Journal of the Soil Science Society of America
Journal of Colloid Science.

A partial list of recent text and reference works on clays is given below:

General works

Barrer, R. M. and Tinker, P. B. (1984) *Clay Minerals: their Structure, Behavior and Use*, Proceedings of the Royal Society, London, 432 pp.

Bowers, T. S., Jackson, K. J. and Helgeson, H. C. (1984) *Equilibrium Activity Diagrams*, Springer Verlag, Berlin, 397 pp.

Garrels, R. M. (1960) *Mineral Equilibria at Low Temperatures and Pressures*, Harper and Row, New York, p 254.

Grim, R. E. (1968) *Clay Mineralogy*, McGraw-Hill, New York, 569 pp.

Newman, A. C. D., ed. (1987) *Chemistry of Clays and Clay Minerals*, Mineralogical Society of G.B., Monograph 6, Longman, London, 480 pp.

Velde, B. (1985) *Clay Minerals: A Physico-chemical Explanation of their Occurrence*, Elsevier, Amsterdam, 427 pp.

Weaver, C. E. (1989) *Clays, Muds and Shales*, Elsevier, Amsterdam, 819 pp.

Methods of investigation and mineralogy

Bailey, S. W., ed. (1988) *Hydrous Phyllosilicates* (exclusive of micas), Reviews in Mineralogy Vol. 19, Mineralogical Society of America, Washington D.C., 725 pp.

Brindley, G. W. and Brown, G. (1980) *Crystal Structure of Clay Minerals and their X-ray Identification*, Mineralogical Society of G.B., Monograph 5, London, 495 pp.

Farmer, V. C. (1974) *Infrared Spectra of Clay Minerals*, Mineralogical Society of G.B., London.

Fripiat, J. J., ed. (1982) *Advanced Techniques for Clay Mineral Analysis*, Elsevier, Amsterdam, 235 pp.

Mackenzie, R. C. (1957) *The Differential Thermal Analysis of Clays*, Mineralogical Society of G.B., London 456 pp.

Moore, D. C. and Reynolds, R. C. (1989) *X-ray Diffraction and the Identification of Clay Minerals*, Oxford University Press, 332 pp.
Sudo, T., Shimoda, H. Yotsumoto, H. and Aita, S. (1981) *Electron Micrographs of Clay Minerals*, Elsevier, Amsterdam, 203 pp.
Wilson, M. J., ed. (1987) *A Handbook of Determinative Methods in Clay Mineralogy*, Blackie, Glasgow, 308 pp.

Clays in soils

Birkland, G. W. (1984) *Soils and Geomorphology*, Oxford University Press, 372 pp.
Bohn, H. L., McNeal, B. L. and O'Connor, G. A. (1985) *Soil Chemistry*, Wiley, New York, 341 pp.
Bolt, G. H. ed. (1982) *Soil Chemistry: Physico-chemical Models*, Elsevier, Amsterdam, 527 pp.
Bolt, G. H. and Bruggenwert, M. G. M. (1978) *Soil Chemistry*, Elsevier, Amsterdam, 281 pp.
Dixon, J. B. and Weed, S. B. eds. (1989) *Minerals in Soil Environments*, Soil Science Society America, Madison, Wisc., 1244 pp.
Fitzpatrick, E. A. (1980) *Soils: their Formation, Classification and Distribution*, Longman, London, 353 pp.
Gieseking, J. E. ed. (1975) *Soil Components*, Springer-Verlag, 684 pp.
Greenland, D. J. and Hayes, M. H. B. (1981) *The Chemistry of Soil Processes*, Wiley, New York, 714 pp.
Marshall, T. S. and Holmes, J. W. (1979) *Soil Physics*, Cambridge University Press, 325 pp.

Clays and the environment

Chapman, N. A. and McKinley, I. G. (1987) *The Geological Disposal of Nuclear Waste*, John Wiley, New York, 280 pp.
Milnes, A. G. (1985) *Geology and Radwaste*, Academic Press, London, 328 pp.
Sly, P. G. ed. (1986) *Sediments and Water Interaction*, Springer-Verlag.
Theng, B. K. G. (1974) *The Chemistry of Clay Organic Reaction*, Adam Hilger, London, 343 pp.
Theng, B. K. G. (1979) *Formation and Properties of Clay-polymer Complexes*, Elsevier, Amsterdam, 362 pp.
Zachar, D. (1982) *Soil Erosion*, Elsevier, Amsterdam, 547 pp.

Industrial applications

Banin, A. and Kafkafi, U. eds. (1980) *Agrochemicals in Soils*, International Irrigation Center, Ottawa, 448 pp.

Barrer, R. M. and Tinker, P. B. (1984) *Clay Minerals, their Structure, Behavior and Use*, Royal Society of G.B., London, 432 pp.
Grim, R. E. (1962) *Applied Clay Mineralogy*, McGraw-Hill, New York, 422 pp.
Grim, R. E. and Guven, N. (1978) *Bentonites: Geology, Mineralogy and Uses*, Elsevier, Amsterdam, 256 pp.
Theng, B. K. G. (1979) *Formation and Properties of Clay-polymer Complexes*, Elsevier, Amsterdam, 362 pp.

3

Clays as minerals

This chapter examines the basic criteria which determine the species of clays most often found in nature. The criteria are of structural (XRD) and chemical types. These criteria and the data on which they are based are necessary for an understanding of the precise differences between the different types of clay mineral. However, it is not necessary to know the details of cell dimensions or octahedral site ion content in order to use or deal with clays. For this reason, a brief description of the clay types follows which provides an overview of the problem of nomenclature in clay mineralogy.

As was stated in Chapter 1, clays are of two major types, swelling and non-swelling. Since all clays are dominated by silica, their SiO_2 content is not of great diagnostic help in their identification. Generally the elements Al, Mg, Fe, K, and to a lesser extent Na and Ca, are useful indicators of clay type. The last important criterion is the distance between the sheet layers of the crystal structures. This is called the **basal spacing**. It is determined after heating to 200°C to eliminate absorbed water and thus called the dried state. Swelled spacings are determined using ethylene glycol vapour to expand the layers to a standard distance.

Taking these three elements as a basis for rapid identification, one can construct a rough table of clay mineralogy (Table 3.1).

Given this very schematic description of clay mineralogy, it is evident that several of the mineral types have different names, different compositional dominants and the same behaviour under XRD examination. It is still evident that there is some purpose in a deeper investigation of the details of clay mineralogy.

It is the custom of clay mineralogists to use simple mineral structures to explain the compositional substitutions and properties of clay minerals. The model structures chosen are often not those of true clay minerals. They are used to indicate the major characteristics of clay chemistry and structural relations.

Table 3.1

	Dominant elements	*Basal spacing (Å)* *Glycol*	*Dry*
SWELLING TYPES			
Smectites			
Beidellite	Al	17	10
Montmorillonite	Al (Mg, Fe^{2+} minor)	17	10
Nontronite	Fe^{3+}	17	10
Saponite	Mg, Al	17	10
Vermiculite	Mg, Fe^{2+}, Al (Fe^{3+} minor)	15.5	10–12
Mixed layer minerals*		10–17	<10
NON-SWELLING TYPES			
Illite	K, Al (Fe, Mg minor)	10	
Glauconite	K, Fe^{2+}, Fe^{3+}	10	
Celadonite	K, Fe^{2+}, Mg, Fe^{3+}, Al^{3+}	10	
Chlorite	Mg, Fe, Al	14	
Berthiérine	Fe^{2+}, Al^{3+} (minor Mg)	7	
Kaolinite	Al	7	
Halloysite	Al	10.2	
Sepiolite	Mg, Al	12.4	
Palygorskite	Mg, Al	10.5	
Talc	Mg, Fe^{2+}	9.6	

*Two or more types of basic layer interstratified in the same crystal

3.1 STRUCTURES

3.1.1 Basic units

Before describing a complete clay structure it is necessary to understand the fundamental molecular units involved in the clay structures and the arrangements of these molecules in the overall patterns common to all clays.

Tetrahedra

Most clay minerals have a special, or at least characteristic arrangement of their constituent atoms in long, interlinked planes. This structure, called a **sheet silicate** or **phyllosilicate** structure, is determined by an essentially two-dimensional cross-linking of SiO_2 units. Each silicon atom is surrounded by four oxygen atoms to form a **tetrahedron** (Fig. 3.1). This is the most basic unit of the clay structures. The tetrahedrally co-ordinated silicon cations are linked one to another by highly covalent bonding through sharing of oxygens (Fig. 3.2). These shared oxygens

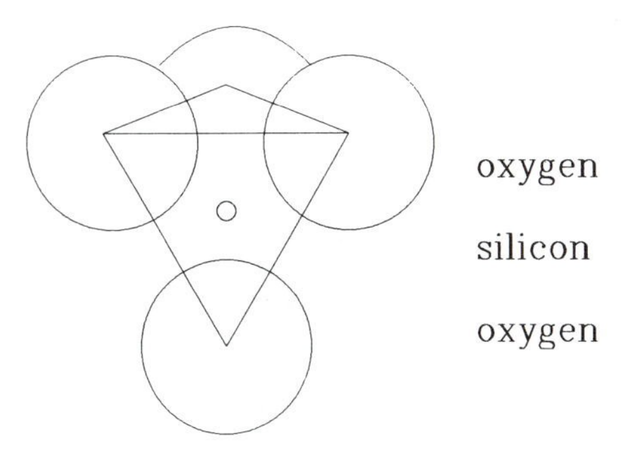

Fig. 3.1 Silicon tetrahedral arrangement with one silicon atom surrounded by four oxygen anions.

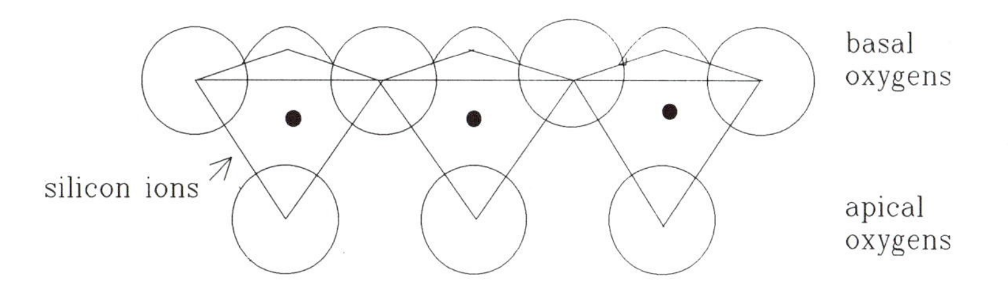

Fig. 3.2 Linked silica tetrahedra showing the shared oxygen ions between two silicon ions.

form a plane of atoms along one base of the tetrahedral structural units. The interlinked oxygens are called **basal oxygens**. The linked tetrahedra form a two-dimensional array of atoms, the basis of the sheet structure.

The arrangement of the interlinked basal oxygens of the tetrahedra occurs in such a way as to leave a hexagonal-shaped 'hole' in the network of oxygen atoms (Fig. 3.3). These holes or cavities are important in processes of attraction between successive basal tetrahedral planes and linking the sheets of the clay structures one to another.

On the opposite end of the linked tetrahedra which form the sheet structure one finds the **apical oxygens** of the tetrahedra, those pointing away from the interlinked tetrahedral bases. These apical oxygens are shared with another series of cations. This distinguishes the apical oxygens from the basal oxygens of the tetrahedra. Basal oxygens are linked in shared tetrahedra only while apical oxygens are shared with cations to form another polyhedron. The tetrahedrally coordinated cations (dominated by Si cations) are thus firmly linked with another layer of cations through their apical oxygens.

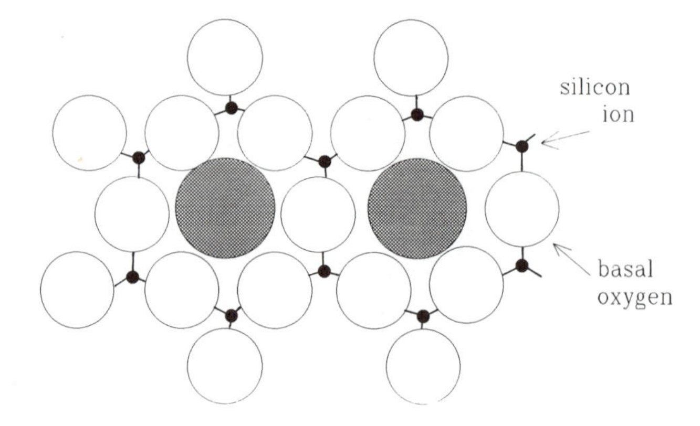

Fig. 3.3 Hexagonal arrangement of the basal oxygens of the linked silica tetrahedra. Dots are the silicon ions (situated below the plane of the surface oxygens), shaded circles show the positions which an interlayer ion will take if the charge on the oxygen surface is sufficiently high to fix it onto the clay structure surface.

Octahedra

In each clay structure there are cations which are coordinated with six oxygens or hydroxyl units in an **octahedral** polyhedron (Fig. 3.4). As in the case of tetrahedra, the octahedrally coordinated cations are interlinked by shared anions in a two-dimensional sheet structure. The octahedral cations are also interlinked with tetahedral sheet cations through shared oxygens, the apical oxygens of the silica tetrahedra. In clay structures there are two directions of oxygen sharing: in the sheet direction, and between the tetrahedral and octahedral units.

Interlayer ions

Cation substitutions in the sheet structure can create charge imbalance on the layer. The charge imbalance is satisfied by inserting cations into the holes in the basal oxygen array. The extra cation effects a balance of charge on the composite structure (that is, two adjacent layers). The two-dimensional sheets, bonded through an interlayer compensating cation, are stacked upon one another in a regular manner to form a three-dimensional crystal. The sheets are held firmly together by cations attracted into the holes in the oxygen array. The diameter of the hole in the basal oxygen array is close to that of potassium, ca. 3 Å.

In fact an ideal hexagonal array of oxygens is rare in clays or phyllosilicates observed at surface conditions (22°C, 1 atm). The cavity tends to be

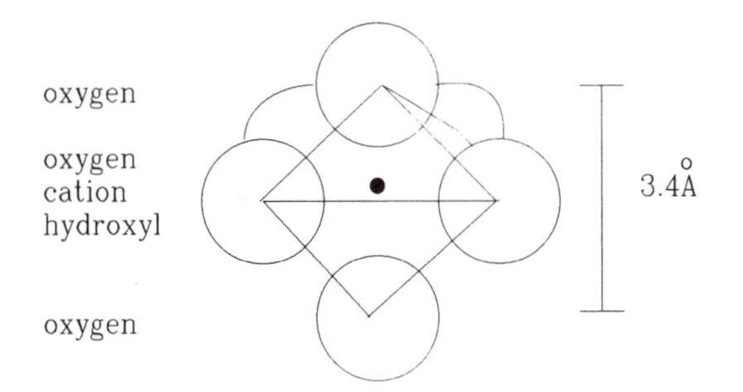

Fig. 3.4 Illustration of the octahedrally coordinated cation polyhedron.

deformed into a ditrigonal shape which is accentuated by the species of cations present in the tetrahedral positions and to a lesser extent the ions present in other sites in the structure. This ditrigonal deformation changes the coordination of oxygen ions in one layer from six to three. This affects the attraction of certain cations for this crystallographic site. The more ditrigonal the sheet, the lower the coordination number of the oxygens surrounding an ion inserted in this site.

Crystallographic orientations

The linking of tetrahedral and octahedrally coordinated cations is the fundamental characteristic of the sheet structure minerals or phyllosilicates. The coordinated polyhedra are linked in a plane, which is the a–b crystallographic direction of the structure. The linking in the a–b plane is considered to be semi-infinite compared to the linking by shared oxygens in the direction perpendicular to the 'sheet'.

The linking of the tetrahedra and octahedra is done in such a manner as to offset the successive oxygen layers slightly. The octahedral ion does not fall just below the tetrahedral ion in the sheets. The mineral structures are then not orthogonal in all directions. They are monoclinic. The monoclinicity is measured by the crystallographic angle due to the offset between tetrahedra and octahedra. The direction perpendicular to the sheet layer is then not simply c but $c \sin \beta$, where β is the monoclinic angle. The (001) repeat distance is the distance between equivalent layers of basal oxygens or oxygen-hydroxyls in a clay structure. It is the $c \sin \beta$ distance. As it is difficult to measure the β angle, clay mineralogists usually use the (001) or basal spacing of the structure to identify and characterize clays.

The crystallographic dimensions within the sheet layers are generally such that $\beta = 3^{1/2}a$. Therefore most crystallographic data reported for

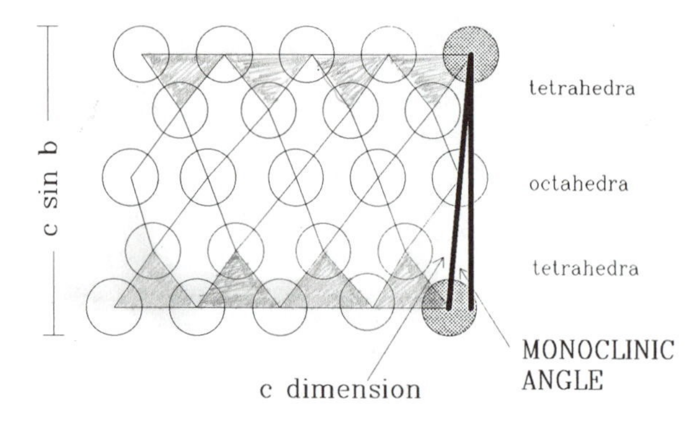

Fig. 3.5 Displacement of the layers of coordinated ions from tetrahedra to octahedra, which gives the monoclinic angle (β). The vertical direction, perpendicular to the ab plane of linked ions is in fact the c sin β dimension. This is the common value, (001) basal spacing, found by XRD measurements of interplanar distances between the *a* and *b* atomic layers.

clays consists of the β cell dimension and the c sin β value which is the (001) basal spacing of the structure (Fig. 3.5).

3.1.2 Chemistry of the coordination units

Tetrahedra

Cations found in the tetrahedrally coordinated sheet are principally Si. Some substitution of Al is common, and occasionally Fe^{3+} is present. It is always assumed by convention that all of the sites in the tetrahedral layer are occupied stoichiometrically. All of the anions in the two fundamental planes are oxygen atoms (Fig. 3.6).

Octahedra

Cations found in the octahedrally coordinated layer are more varied in species: Al, Mg, Fe^{2+} are the principal species, but Fe^{3+}, Ti, Ni, Zn, Cr and Mn can also be present. One of the apical oxygen anions is shared with the tetrahedral units. The intermediate, median, layer of cations of the octahedra consists not only of oxygen but also of hydroxyl anion groups which are shared with other octahedrally coordinated ions to form the sheet structure. The hydroxyl anionic units can be replaced to a certain extent in some clay structures, by F or Cl anions. The apical oxygens of the octahedral polyhedra, when not coordinated with tetrahedra, are made up uniquely of OH units.

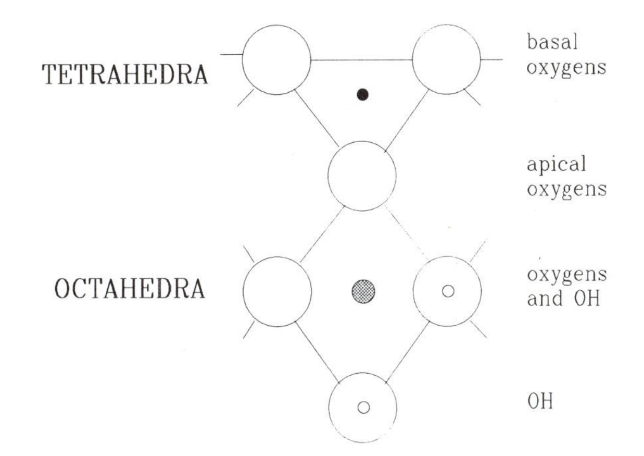

Fig. 3.6 Linkage between tetrahedra and octahedra through a common oxygen anion. Octahedra contain oxygen and hydroxyl anions whereas tetrahedra contain only oxygen anions.

The polyhedral anions are as follows: oxygen for tetrahedra; oxygen for linking cations between tetrahedra and octahedra; OH and oxygen linking octahedra to one another; OH when no linking occurs.

Interlayer ions

The charge compensating ions found between the layers of basal tetrahedral oxygens are of two sorts: those tightly fixed in place, and those which can be easily exchanged. In clays, the tightly fixed interlayer cations are almost exclusively potassic. The exchangeable or **exchange** ions are more varied. They can be monovalent or divalent. They are normally surrounded by water molecules. Virtually any hydrated cation can be found in the exchange site of clays. In natural clays calcium is predominant, while sodium and magnesium are common.

3.1.3 Cell dimensions and ionic substitutions

If there is a substitution of one cation species for another in either the tetrahedral or octahedral layer, the overall dimensions of the coordination polyhedra are changed. The average effective diameters of the different ions are known for the two coordination polyhedra in the phyllosilicates (and other silicates). These diameters are given in Table 3.2.

The substitution of one ion for another will affect the cell dimensions of the clay mineral. However, the relation between cell size and the cumulative diameters of the ions is often not symmetric. For example, when Al

Table 3.2 Radii of common clay-forming cations

Ion	*Radius (Å)*
Tetrahedra	
Si^{3+}	0.26
Al^{3+}	0.39
Fe^{3+}	0.52
Octahedra	
Al^{3+}	0.39
Fe^{3+}	0.63
Mg^{2+}	0.72
Fe^{2+}	0.92
Ni^{2+}	0.66

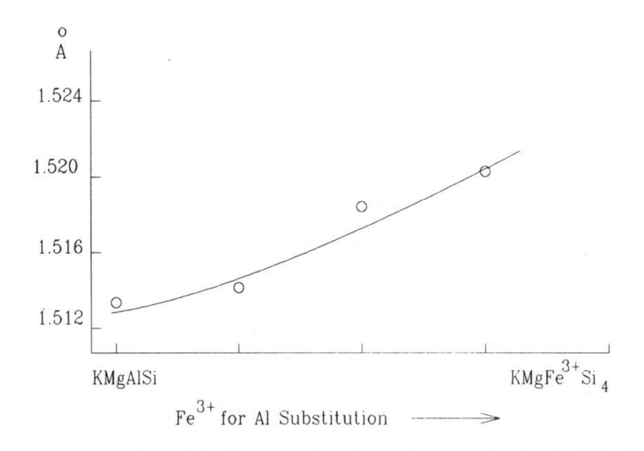

Fig. 3.7 Relation between the types of ions in the octahedral site of celadonite and the (006) cell dimension. Fe^{3+} has a larger diameter than Al^{3+}.

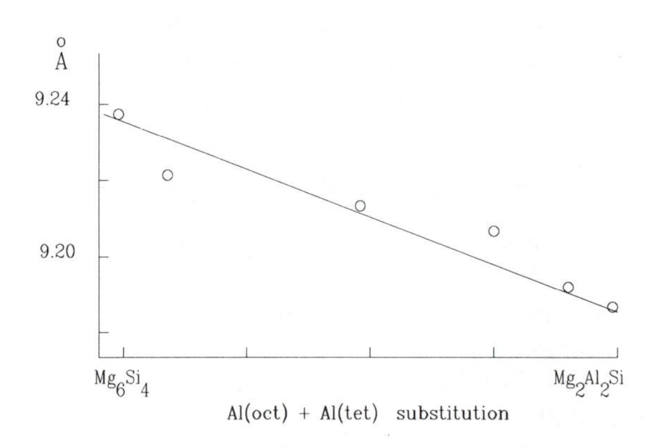

Fig. 3.8 Effect of ionic substitution in chlorites where Al is substituted in both the octahedral and tetrahedral sites. The overall effect is to decrease the *b* dimension.

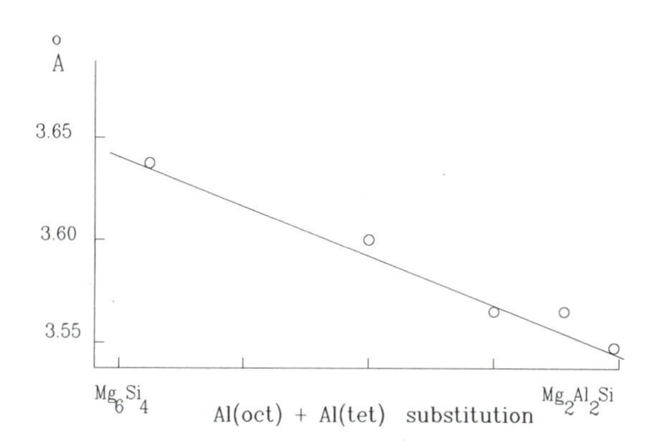

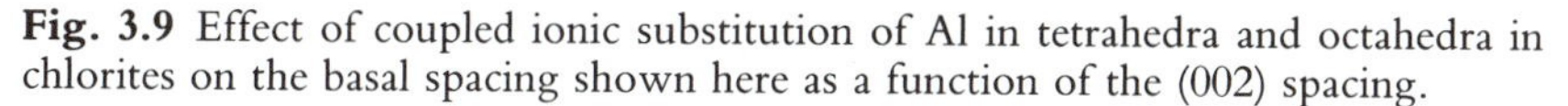
Fig. 3.9 Effect of coupled ionic substitution of Al in tetrahedra and octahedra in chlorites on the basal spacing shown here as a function of the (002) spacing.

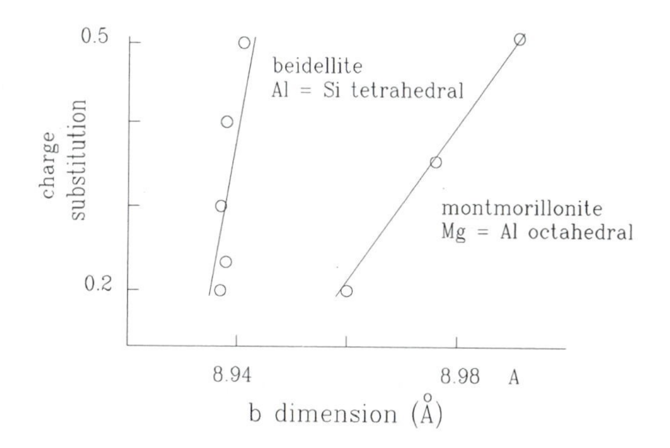

Fig. 3.10 Effect of ionic substitution in the tetrahedral site (beidellites) and octahedral sites (montmorillonites) of dioctahedral smectites. Charge substitution increases as an increasing number of ions are substituted into the structure.

is substituted for Si in the tetrahedral site, the change in the mineral cell dimension (increase) is greater in the *c* (or *c* sin β) direction than in the *a* and *b* directions. Substitution of Fe for Mg in the octahedral site greatly increases the *b* and *a* dimensions but only slightly that of the *c* (*c* sin β) dimension. Some examples of substitutions and change in cell dimension are given in Figs 3.7–3.10.

The substitution of Fe^{3+} for Al in synthetic celadonite (potassium, iron-rich mica) shows an increase in the *b* cell dimension (Fig. 3.7) but it is not as great as would be expected in comparing the ionic radii. Three-quarters substitution should give a change of 0.27 Å whereas one sees a difference of only 0.08 Å in the synthetic celadonite minerals. Another example is that of the effect of Al substitution in both the octahedral site (Al = Mg) and tetrahedal site (Al = Si) in synthetic chlorites. In the tetrahedral site the Al ion is larger than the Si ion while in the octahedral site the Al ion is significantly smaller than the Mg ion. The overall change in cell dimension is one of a decrease in both the b (sheet direction) (Fig. 3.8) and $c \sin \beta$ direction. The increase in the ionic dimension due to the tetrahedral substitution is only half of the decrease in size of the Mg–Al substitution. A third example is given in the smectite series with a change in the b dimension due to substitution in the tetrahedra (Al = Si) in the beidellites. This substitution influences the b dimension only slightly whereas the substitution of Mg for Al (octahedral substitution) in montmorillonites changes the cell dimension much more (Fig. 3.10).

3.1.4 Charge relations of polyhedra

As stated above, each cation in the layer silicates is surrounded and coordinated to a group of oxygen ions or hydroxyl ions. This indicates that the cations must balance a given number of negative charges, two for oxygen ions and one for hydroxyls.

The negative charge on a layer silicate unit cell of the tetrahedrally coordinated ions is equivalent to four oxygen ions or eight charges. The basic cation content of a unit cell layer of tetrahedrally coordinated ions is given as $2Si^{4+}O_4$.

The charge on the unit cell of the octahedrally coordinated layer is 6 or that of three oxygen ions. The basic content of such a unit cell is given as $xR^{n}O_3$, where R is the cation, which can be either divalent ($n = 2$) or trivalent ($n = 3$), with $x = 6/n$. In each octahedrally coordinated layer, a unit cell will have six positive charges; however, one can find either di- or trivalent ions present in the majority of sites. This leads to two types of occupancy of the octahedrally coordinated sites, one with three divalent ions ($3R^{2+}$) or one with two trivalent ions ($2R^{3+}$). Thus we speak of **trioctahedral** (three-ion) site occupancy and **dioctahedral** (two-ion) site occupancy of the octahedrally coordinated layer of cations. The charge total of 6 is almost always maintained in clay minerals.

3.1.5 Substitutions in coordination sites

The varieties or species of clays found in nature are formed through substitutions of cations of the same or different charges in the tetrahedral and octahedral sites. These substitutions determine the mineral name given by mineralogists, determine the physical properties of the particles and

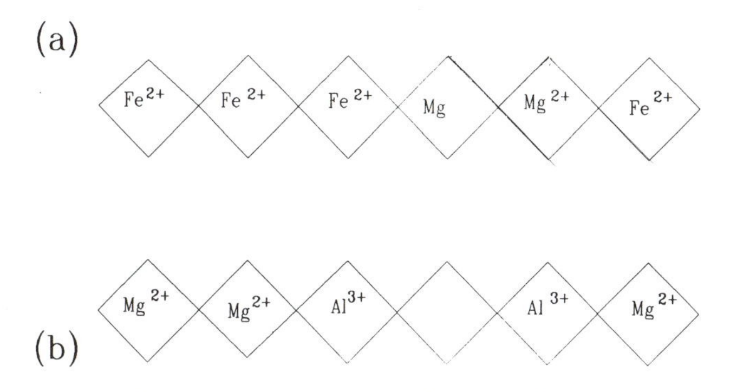

Fig. 3.11 (a) Iso-charge substitution of Mg^{2+} for Fe^{2+}; (b) non-stoichiometric substitution of Al^{3+} for Mg^{2+} in octahedral sites.

determine the stability limits of the clay minerals as a function of thermal and chemical parameters.

Octahedral substitutions

Equivalent charge substitutions are those of the same charge per cation. A continuous sequence of composition can be formed without any change in another coordination site. For example, the classical substitution of Mg^{2+} for Fe^{2+} can be made in the octahedral site alone. Also one can find the Fe^{3+} for Al^{3+} substitution.

Another possibility is a change in the total number of ions present in the octahedral site (in all mineral structural calculations it is assumed that the tetrahedral site is always filled to a capacity of two cations per layer). In the cases of **non-stoichiometric** substitution in the octahedral site, the charge balance is maintained by a substitution of the type $3R^{2+} = 2R^{3+}$, or, in the case of a specific, partial substitution, $3Mg^{2+} = 1.5Mg^{2+}Al^{3+}$ (see Fig. 3.11).

In the substitution as written, the average site occupancy of the octahedrally coordinated layer changes from three ions to two and a half. The octahedral occupancy in clay minerals and other phyllosilicates has been observed to lie between 2.5 and 3 ions for some minerals and to be very near 2 ions for the others. Two groups of minerals can be distinguished those of di-octahedral nature and those of tri-octahedral nature. There is a gap in the substitution series between an ionic occupancy of two-and-a-half and two ions, thus defining the limits between di- and trioctahedral minerals.

The above substitution, called **di-trioctahedral**, reduces the average number of sites occupied in the octahedrally coordinated layer from three

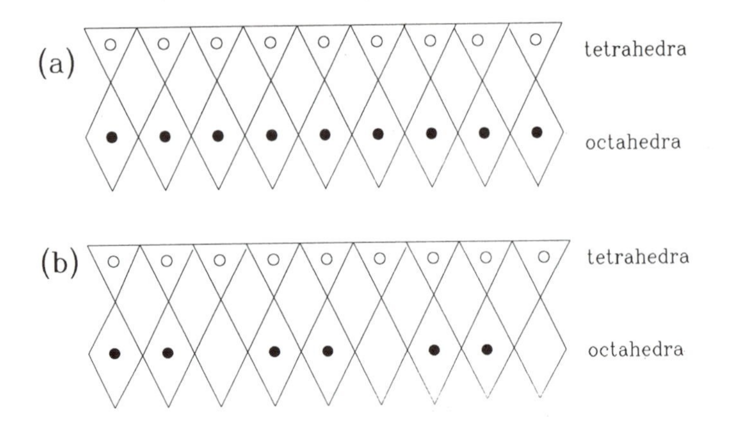

Fig. 3.12 (a) Trioctahedral and (b) dioctahedral structures due to filling of the octahedral sites by three or two cations per three sites.

to less than three. Since a half-ion cannot exist easily in nature, this means that some of the triad octahedral units will contain Mg^{2+}/Al^{3+} and others $Mg^{2+}/Mg^{2+}/Al^{3+}$ since it is not possible to fill a site with a half-atom. In the di-trioctahedral substitution, some triads contain three and others two cations, where all substitutions occur in the same octahedrally coordinated ionic layer (Fig. 3.12).

It is apparent that the *b* cell dimension is much more sensitive to ionic substitution and as a result it is more often used to determine the approximate composition of the octahedral site of a clay mineral. The alumina content of the tetrahedral site can be estimated by observing the $c \sin \beta$ dimension (basal spacing) in the absence of other substitutions. The small ion diameters of the trivalent ions in the two-site occupancy of a dioctahedral mineral diminish the *b* dimension compared to a large ion, three- ion occupancy in a trioctahedral mineral. Therefore, a rapid test to determine the type of mineral structural type (di- or trioctahedral) can be made using the (060) reflection as a measure of the *b* dimension. Dioctahedral minerals are found to have a reflection near 1.50 Å and trioctahedral minerals one near 1.54 Å.

Tetrahedral–octahedral substitutions

If the ion substituted in the octahedral site is of different charge – for example, Al^{3+} for Mg^{2+} – and all sites in the structure are filled, a substitutional charge compensation must be made in another cation site in the structure in order to maintain electrostatic neutrality on the layer structure. In the above situation one can conceive of a coupled substitution

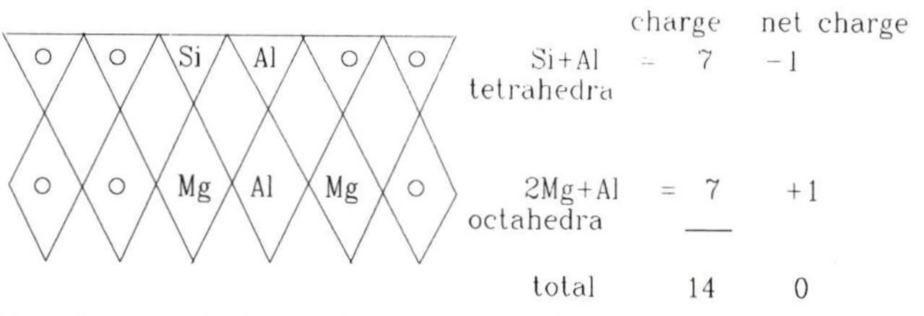

Fig. 3.13 Coupled tetrahedral and octahedral substitutions giving an electronically equilibrated structure.

in tetrahedral and octahedral units, where Al^{3+} replaces Mg^{2+} with a gain of one charge in the octahedral site and Al^{3+} replaces Si^{4+} in the tetrahedral site with a corresponding loss of one charge. In this way electrostatic neutrality is maintained for the same total number of anions in the structure. Both the tetrahedral and octahedrally coordinated layers are involved in the substitutions. This can be written as

$$(3R^{2+})\text{oct}(2Si^{4+})\text{tet} = (2R^{2+}R^{3+})\text{oct}(Si^{4+}R^{3+})\text{tet}$$

where the total charge of octahedral plus tetrahedral layer remains 14 (see Fig. 3.13).

Interlayer substitution

The types of substitution mentioned above involve only the ions in the adjacent, oxygen-sharing tetrahedral and octahedral units. It is possible to have an ion substitution which is electrostatically balanced by the addition of another ion on the surface of the layer sheet of the tetrahedral-octahedral network: the interlayer ion substitution. The substitutions which create a charge imbalance on the sheet occur only in structures with two tetrahedrally coordinated and one octahedrally coordinated layer.

Three types of substitution are possible. One is where a trivalent ion (R^{3+}) is substituted for Si^{4+} in the tetrahedral site, creating a charge balance of −1 on the sheet structure. This deficit is compensated by an ion, held by dominantly ionic bonding on the oxygen surface of the sheet structure between two tetrahedral layers of adjacent sheet structures (Fig. 3.14). Normally the ion has a charge of +1 or +2, with Na, K, Ca, Mg the most common ions present.

Another substitution giving rise to the addition of a compensating ion on the oxygen surface interlayer site is the substitution of a divalent for a trivalent ion in the octahedrally coordinated layer, such as Mg^{2+} for Al^{3+}, (Fig. 3.15).

A third, rarer, substitution is by a decrease in the octahedral site

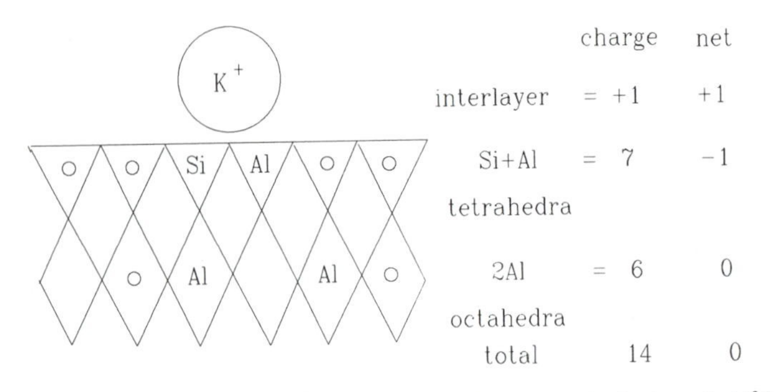

Fig. 3.14 Non-equilibrated substitution in the tetrahedral site of Al^{3+} for Si^{4+} giving rise to a net charge imbalance of 1^+ which is compensated by an interlayer cation (potassium) to give a globally equilibrated structure. This substitutional scheme is called a **mica** (dioctahedral in the example given).

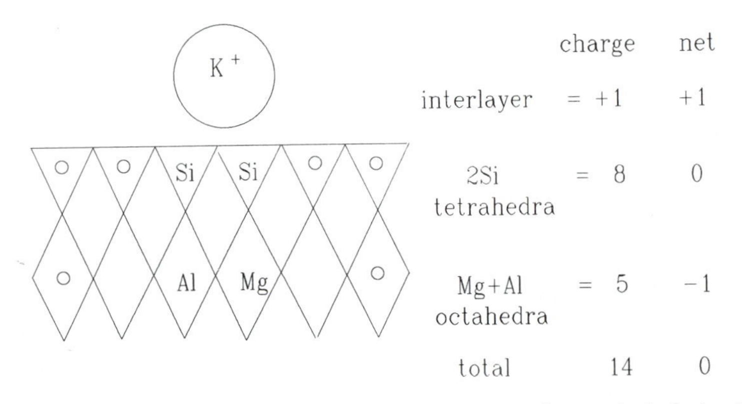

Fig. 3.15 Non-equilibrated substitution originating in the octahedral site (Mg^{2+} for Al^{3+}) compensated by a potassium interlayer ion. The electronically equilibrated structure is a dioctahedral mica, like the one in Fig. 3.14.

occupancy (non-stoichiometric), which decreases the overall charge on the structure creating the need for an interlayer ion. Of the normal three-ion sites in the structure, there are, then, two types of octahedral site occupancy, di- and trioctahedral, but in all cases the octahedrally co-ordinated ions present are only divalent (R^{2+}). For every six octahedral sites only five will be occupied (Fig. 3.16).

The substitutions which necessitate the addition of a compensating interlayer ion are as follows: for the tetrahedral case:

$$(2R^{3+})\text{oct}(4Si^{4+})\text{tet} = (2R^{3+})\text{oct}(3Si^{4+}, R^{3+})M^{+}$$

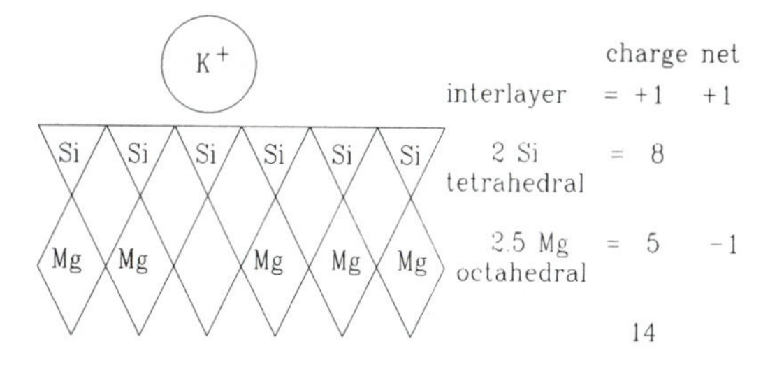

Fig. 3.16 Vacancy 'substitution' which creates a net charge compensated by an interlayer potassium cation. This gives rise to a **tetrasilic mica substitution**.

where the substitution occurs in the tetrahedral and interlayer site (M^+); for the octahedral case:

$$(2R^{3+})oct(4Si^{4+})tet = (R^{3+}R^{2+})oct(4Si^{4+})M^+$$

where the substitution is made in the octahedral and interlayer site; and for the octahedral vacancy case:

$$(6R^{2+})oct(8Si^{4+})tet = (5R^{2+})oct(8Si^{4+})2M^+$$

where a di- and trioctahedral site triad exists in the mineral layers which is compensated by two interlayer ions.

3.2 POLYTYPES

The clay structures are layers of oxygen-coordinated cations which are interlinked into two-dimensional sheets. These sheets are stacked one on top of the other to form a three-dimensional crystal. In each sheet there is an offset between the tetrahedral and octahedrally coordinated ions, forming a monoclinic angle. This means that all of the interlinked tetrahedral basal oxygens and those of the other polyhedral sites will be equidistant in the (001) or $c \sin \beta$ direction. However, there will not be the same regularity in atomic arrangement in the *a* and *b* crystallographic directions. The monoclinic offset places the different ions in more specific crystallographic sites. The result is that the dominant X-ray crystallographic reflections of the ions in clays is (001) where many atoms are aligned to form the basal planes. The other primary reflections, such as (010) or (100), contain fewer atoms and are not as strong in the X-ray diffractogram.

A second effect of the monoclinic displacement is that as the sheets are stacked one on top of the other, the relative orientation of the sheets with respect to one another will give different diffraction planes. Different stacking arrangements give different X-ray diffraction patterns called

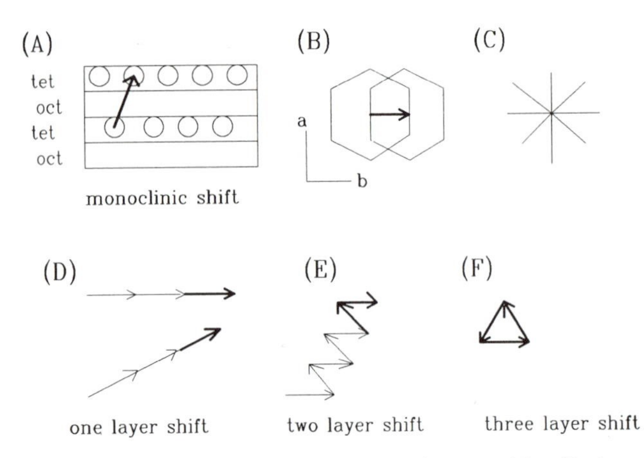

Fig. 3.17 Stacking sequences which give rise to polytypes. The displacement of an atomic position from one layer of a structure to another is due to the monoclinic shift from mineral layer to mineral layer as the crystal is formed. The orientation of the monoclinic shift in going from one layer to the next produces different crystallographic structures based upon more than one basic sheet unit, that is, a one-layer shift, a two-layer shift, a three-layer shift, etc. The shifts are guided by the hexagonal symmetry of the oxygen basal plane ions.

polytypes. If we consider the direction of the monoclinic shift from one unit layer of the structure to another, we can follow the variations in stacking arrangement. Figure 3.17 illustrates this polytypism for a two-layer structure. As the basic structural motif in clays is a hexagon, the oxygens in the basal layer of the tetrahedral sheet form a hexagonal array. The basal oxygen array is determinant in the stacking motif as one sheet is superposed on another. The hexagonal array determines the amount and direction of displacement from one layer to another in the three-dimensional crystal.

For example, take two hexagonal arrays of tetrahedral basal oxygens from successive layers in a clay crystal (Fig. 3.17b). In profile, perpendicular to the a–b crystallographic plane, the monoclinic shift is seen as an inclination of the vector between two equivalent crystallographic sites (Fig. 3.17a). However, seen as a projection on the a–b plane, the shift can have a vector in the a–b plane which will be determined by the hexagonality of the oxygen layers, that is, in directions inclined 60° or 180° to one another (Fig. 3.17c). If the stacking sequence produces monoclinic shifts in the same direction it will form a one-layer polytype (Fig. 3.17d). If the shift repeats itself every two layers (Fig. 3.17e) the result is a two-layer polytype. If the shift has a three-layer periodicity one will have a three-layer polytype (Fig. 3.17f).

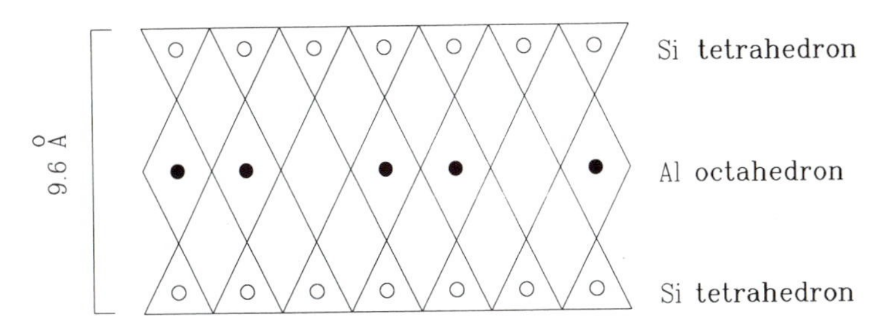

Fig. 3.18 Pyrophyllite, with a neutral dioctahedral 2:1 structure.

In all cases the basic sheet structure is crystallographically the same, and has the same composition. The polytypes have (001) reflections in common but not (hkl) reflections. The names or identifying symbols of the different polytypes are usually based upon the number of basic sheet layers involved in the shift cycles. In the following descriptions of mineral groups, XRD data on polytypes will be given when they have been used to identify specific clay structures to geological advantage, that is, designating physical conditions of formation or chemical composition of a clay type.

3.3 CLAY MINERAL STRUCTURES

In order to describe the different major types of ionic arrangements in the basic structural types, several model structures are given as a basis for the derivation of the more complicated clay mineral structures.

3.3.1 2:1 structures

Pyrophyllite (dioctahedral)

This structure is used to describe the mica and dioctahedral smectite mineral groups. The basic coordination units are two tetrahedra and one octahedron. The octahedral unit is coordinated through a shared or linking oxygen to two layers of tetrahedrally coordinated ions. Hydroxyls are found only in the intermediate layer of the octahedrally coordinated structural unit (Fig. 3.18).

In the mineral pyrophyllite, only Si is found in the tetrahedral layer sites and almost exclusively Al is present in the octahedral sites. Some minor Fe^{3+} can also be found in these sites.

This structure of two tetrahedra and one octahedron is called a dioctahedral 2:1 structure. In the octahedral site only two ions fill the three possible sites with a total of six positive charges. The mineral formula is $Al_2Si_4O_{10}(OH)_2$, based upon 22 negative charges from ten oxygens and two OH units. The two hydroxyl units are present in the middle plane of cations of the octahedral layer. These hydroxyls are coordinated only to the trivalent, aluminium ions. Pyrophyllite has the following cell dimensions:

$$c \sin \beta = 9.20\,\text{Å}$$
$$b = 8.96\,\text{Å}$$

The pyrophyllite structure is used as a model for minerals of the dioctahedral 2:1 type. These clay minerals can be derived from the pyrophyllite structure by substitutions of ions in different structural sites. Substitutions are of two types: those which conserve charge on the 2:1 layer (neutral structure); and those which change the cationic charge on the 2:1 layer (charged structure).

Neutral substitution:
octahedral (e.g. $Al^{3+} = Fe^{3+}$)

Charged structure:
interlayer-octahedral (e.g. $Al^{3+} = Mg^{2+(oct)}M^{+(int)}$)
interlayer-tetrahedral (e.g. $Si^{4+} = Al^{3+(tet)}M^{+(int)}$)

Interlayer substitutions are necessary when a low-charge ion is substituted in a site for a high-charge ion. This creates a charge imbalance on the 2:1 structure which is compensated by inserting an ion between the layers of the 2:1 sheets. These interlayer ion substitutions can range from 0.2 to 1.0 charges per unit cell. High-charge substitutions (0.9 to 1.0) fix the layers of the sheet structure together in a tightly bonded structure. This is the mica structure. If the charge is less than one ionic charge the mineral is often referred to as being mica-like. In the clay minerals, only potassium is found in the interlayer position of high-charge structures.

When the charge on the layers is below 0.9, the structure can accept polar molecules between the layers and thus change its interlayer dimension. These are the swelling clays. In the swelling clays Na, K, Ca, Mg are the major interlayer ions.

In the dioctahedral 2:1 minerals both high- (0.9 to 1.0) and low- (0.2 to 0.9) charge minerals can be formed.

Talc

This structure is similar to that of pyrophyllite. However, the occupancy of the octahedrally coordinated layer is complete, three ions are normally present and it is thus **trioctahedral** (Fig. 3.19). The octahedral site can

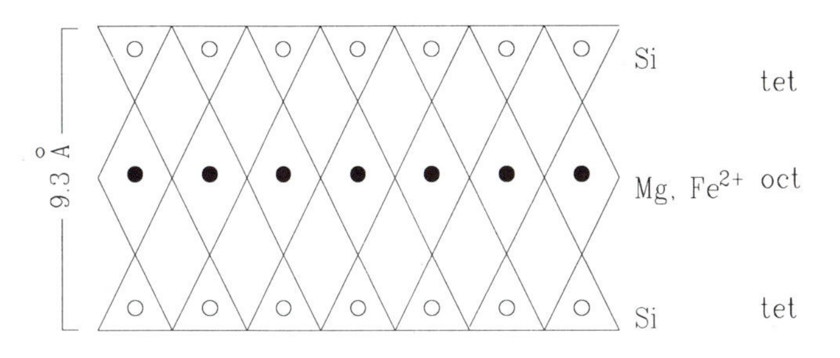

Fig. 3.19 Talc, a 2:1 trioctahedral structure.

show the full range of Mg–Fe substitutions, all ions having the same charge. Talcs are found in metamorphic rocks; they are usually entirely magnesian. The talcs found in soils and weathering profiles have the total range of Mg–Fe substitutions. Therefore this mineral is found in a wide range of thermal conditions. The cell dimensions of Mg talc are as follows:

$$c \sin \beta \ (001) = 9.33\ \text{Å}$$
$$b = 9.14\ \text{Å}$$

Using talc as a model structure, substitutions occur in the tetrahedral and octahedral site as in the pyrophyllite structure giving a neutral or charged layer structure. However, the charged structure, interlayer substitutions are limited to values below about 0.9 charges giving swelling clays and hence there are no mica or mica-like trioctahedral clay minerals:

Neutral structure:
octahedral (e.g. $Al^{3+} = Fe^{3+}$, $Mg = Fe^{2+}$)
di-trioctahedral (e.g. $3R^{2+} = 2R^{3+}$)
octahedral-tetrahedral (e.g. $R^{2+}Si^{4+} = (R^{3+(oct)}R^{3+(tet)})$)

Charged structure:
interlayer-tetrahedral (e.g. $Si^{4+} = R^{3+}K^{+(int)}$)
interlayer-octahedral (e.g., $R^{3+} = R^{2+}K^{+(int)}$)
interlayer-octahedral vacancy ($6R^{2+(oct)} = 5R^{2+(oct)}K^{+(int)}$)

It is evident that there are more substitutional types in the trioctahedral 2:1 structure than in the dioctahedral 2:1 structure. The trioctahedral minerals have much more latitude in substitutional combinations.

In the neutral substitutions, there are the homoionic types, those ions of the same charge. The di-trioctahedral substitution is simply that where two trivalent ions substitute for three divalent ions, the total of six cationic charges in the octahedral sites is conserved. This substitution is only partial, going from a total of three to two and a half ions in the octahedral site.

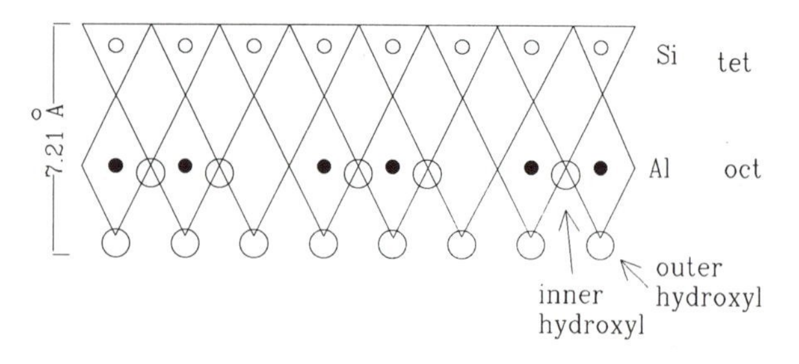

Fig. 3.20 1:1 dioctahedral structure of kaolinite. Two hydroxyl sites are present, one in the octahedral layer on the same level as the aluminium ion and the other at the octahedral surface.

Layer charged substitutions usually involve changing a low-charge ion for a high-charge ion, Al^{3+} for Si^{4+} for example. An additional interlayer ion is present to compensate for the charge imbalance on the structure as a whole. Another substitution is possible, that of decreasing the octahedral site occupancy (non-stoichiometric substitution) in the octahedral site and compensating the charge lost by adding an interlayer ion.

Of course, a single clay mineral can exhibit combinations of these substitutions.

3.3.2 1:1 structures

The mineral structures used for the 1:1 structures, that is, those with one tetrahedral layer and one octahedral layer, are kaolinite and serpentine.

Kaolinite

Kaolinite ($Al_2Si_2O_{10}(OH_6)$) has a simple structure with a single layer of tetrahedrally coordinated ions linked through oxygens to a layer of aluminium octahedra (Fig. 3.20). The summits of the octahedra are exclusively hydroxyl units. The intermediate anion sites are occupied by both oxygens and hydroxyls. We distinguish **inner hydroxyls**, those similar to sites in 2:1 minerals, and **outer hydroxyls**, which are not found in 2:1 structures. The cell dimensions of kaolinite are:

$$(001) = 7.21\,\text{Å}$$
$$b = 8.99\,\text{Å}$$

Serpentine

The trioctahedral 1:1 structure is very similar to that of kaolinite, except that the octahedral site is filled with divalent ions such as $3Mg^{2+}$. This is

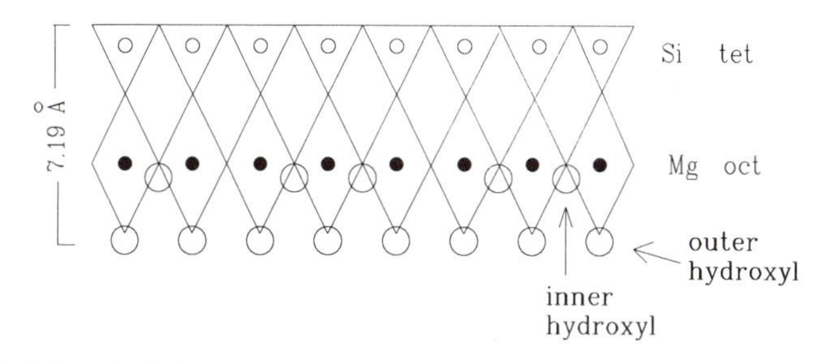

Fig. 3.21 Trioctahedral 1:1 structure (serpentine).

the structure of the serpentine minerals (Fig. 3.21). Both 1:1 structures have a single layer of octahedral cations linked to a layer of silica tetrahedra. The apical, unshared anions on the octahedra are exclusively OH units. Thus the contact between the layers of the 1:1 structure is between basal tetrahedral shared oxygens and apical OH units of the octahedra such as is the case in the 2:1 + 1 chlorite structure (see below).

While the kaolinite (dioctahedral) structure has almost no substitutions in it, it is almost pure Al and Si, the serpentine structure (trioctahedral) has many more possibilities. There can be coupled substitution of trivalent ions in the octahedral and tetrahedral sites ($3Mg^{2+}2Si^{4+} = 2Mg^{2+}2Al^{3+}Si^{4+}$). There is also the possibility of a substitution of the di-trioctahedral type uniquely in the octahedrally coordinated site as $3Mg^{2+} = 1.5Mg^{2+}Al^{3+}$.

The serpentine minerals are high-temperature in origin but the substitutions in the structure can be found in low-temperature clay minerals.

The serpentine mineral family is one formed under conditions of low-grade metamorphism. Thus by its grain size and thermal stability it is not a clay mineral group. The basic interlayer repeat distance is near 7 Å for serpentines; their cell dimensions are:

$$(001) = 7.19\,\text{Å}$$
$$b = 9.25\,\text{Å}$$

3.3.3 2:1 + 1 structures (chlorites)

In chlorites the mineral is built of a trioctahedral 2:1 type structure such as that of talc. There is a second element in the structure which is an octahedrally coordinated layer of ions in which neither of the apical anions is linked to an adjacent tetrahedral layer (see Fig. 3.22). This octahedrally coordinated layer is composed of cations and OH anionic units. The tetrahedral basal oxygens are faced with hydroxyls. The 'interlayer'

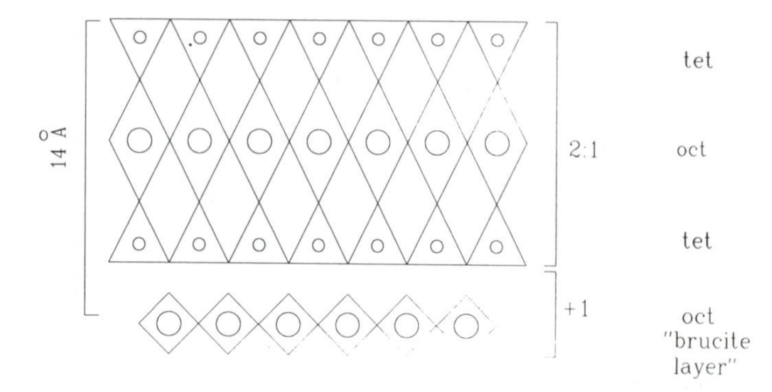

Fig. 3.22 2:1 + 1 chlorite structure composed of linked tetrahedra (two layers) and octahedra. The brucite-type layer is not interlinked with the adjacent tetrahedra but forms an independent unit layer. It occurs in the interlayer site.

octahedral layer is called the **brucite** layer because its bulk composition is similar to that of the magnesium hydroxide mineral, $Mg_3(OH)_6$.

In both of the octahedrally coordinated layers one can find a full range of divalent ion substitutions, $Mg^{2+} = Fe^{2+}$, as well as the di-trioctahedral type of substitution. Further, the coupled octahedral–tetrahedral $3Mg^{2+}2Si^{4+} = 2Mg^{2+}Al^{3+}(Si^{4+}Al^{3+})$ substitution is often observed. The chlorites show a full range of substitutional types, excluding the interlayer type substitution.

The overall thickness of the 2:1 + 1 structure is slightly greater than 14 Å, that of two 1:1 structures laid one on top of the other. However, the structure is significantly different in that there is no linking coordination of apical anions of the octahedrally coordinated unit layers to the basal oxygens of adjacent tetrahedral layers. Magnesian chlorites have the following cell dimensions:

$$c \sin \beta\text{: } 14.05\text{–}14.38\ \text{Å}$$
$$b\text{: } 9.20\text{–}9.25\ \text{Å}$$

3.3.4 Generalizations

One can classify the structures presented above into three major categories:

1. **Neutral lattice structures** (talc, pyrophyllite, kaolinite, serpentine, chlorite). In these structures the 2:1 or 1:1 units of interlinked tetrahedra and octahedra have a net charge of zero. The substitutions within the sheets are cancelled electrostatically and the individual layers (2:1, 1:1 and 2:1 + 1) are bound into a crystal by low-energy bonds (van der Waals type). Talc,

pyrophyllite (2:1), kaolinite and serpentine (1:1) and chlorite (2:1 + 1) are neutral lattice structures.

2. **High-charge mica structures** (0.9–1.0 charge). In these minerals, there is a charge imbalance due to ionic substitution on the basic 2:1 structures which is compensated by an interlayer ion which is fixed between the layers bonding them together. The charge on the 2:1 layer is near 1. The interlayer ion is firmly held between the adjacent layers and it is an integral part of the structure. The layers are firmly bonded to each other. In clays this interlayer ion is almost uniquely potassium. The 2:1 structure is only dioctahedral. There are no 1:1 structure micas.
3. **Low-charge 2:1 structures** (0.2–0.9 charge) where there is a net charge imbalance on the tetrahedral-octahedral network of 0.2–0.9, compensated by loosely held ions in the interlayer position which can be easily exchanged in aqueous solution. These minerals swell (incorporate polar ions between the layers) accepting various types of molecules between the 2:1 layers which are attracted by the electrostatic charge. The low charge does not fix the cations stably between the layers.

Therefore one can speak of neutral charge layer minerals, of micas, and of swelling, low-charge minerals which have the general name of smectites. These three structural types have specific physical properties which are extremely important to their use in industrial processes and in natural biological processes as well as in their effect on the physical properties of the materials in which they are found.

The neutral layer minerals have a tendency to cleave along the layer sheets, giving a lubricating effect when tangential pressure is applied to them (for example, talc). The micas have some of these properties but they are generally more stable mechanically. The smectites can change their cell dimensions by changing the amount of polar molecules which are accepted between the layers. These are the swelling properties. For example, smectites can have a 10 Å basal spacing in an anhydrous environment, but will swell to 15 Å in humid conditions and to 17 Å in wet conditions.

The structural types can be further classified upon the octahedral ion site occupancy, that is, as either di- or trioctahedral.

Expanding and non-expanding clays

The physical properties of the clay minerals are governed to a large extent by their mineralogical composition and structure. As mentioned above, there are three types of minerals. Those of a neutral charge on the basic

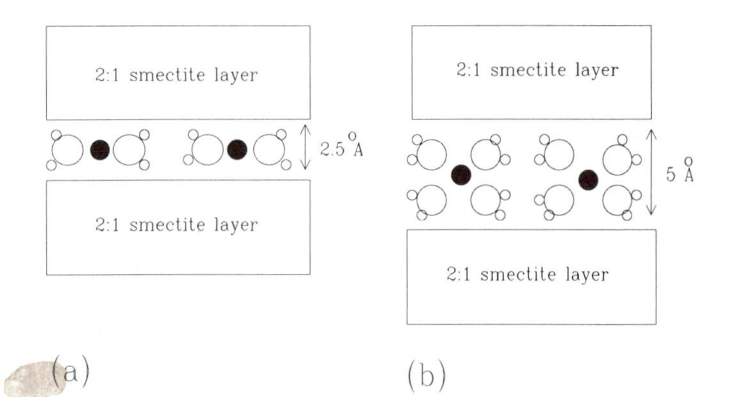

Fig. 3.23 (a) One-layer and (b) two-layer water structure coordinated with the interlayer exchange ion which is found in smectites.

sheet structure, such as kaolinite, talc and pyrophyllite, have no special electronic or chemical affinities with other, non-clay chemical species. The micas or mica-like minerals are also relatively inactive chemically, due to a relatively tightly bonded structure where 2:1 units are interattracted through an interlayer cation due to a high charge density, 1.0 per unit cell. The chemical activity of micas and neutral structures is similar.

However, when the charge imbalance on the 2:1 structure (similar to that for a mica) is less than about 0.9 and greater than 0.2 charges per unit cell, the interlayer cation is less firmly fixed in its site and it can be replaced by another ion (ion exchange) and other polar or ionic molecules can be added into the interlayer structural site. The cations present in these loosely held sites are hydrated under conditions of some humidity and temperatures below 100°C. Depending upon the chemical characteristics of the interlayer cation (charge to diameter ratio) the ion will be hydrated with three or six water molecules. A six-molecule cation gives a two-layer interlayer water structure, expanding the 9.6 Å 2:1 unit layer repeat distance to near 15 Å. A three-molecule hydrated cation will give a structure of one water layer which expands the interlayer spacing of the structure to 12.5 Å (Fig. 3.23). The different exchange cations in their hydrated state have slightly different dimensions of the cation-water complex creating slightly different interlayer spacings in the smectite structures. Monovalent ions give interlayer spacings of slightly less than 15 Å and divalent ions spacings near 15.2 Å.

If the smectite clay mineral is exposed to an atmosphere of another polar ion (liquid or vapour) the molecules can often be inserted around the cation-water complexes, expanding the interlayer spacing still further. A typical, and standard, polar molecule used to measure the expandability of

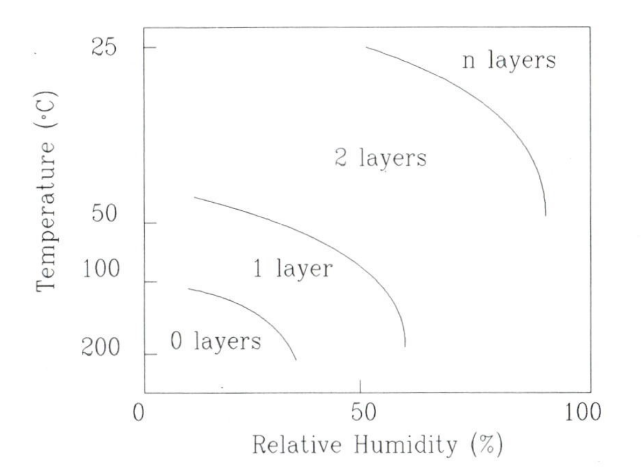

Fig. 3.24 Relationship of the effect of relative humidity and temperature on the number of water layers coordinated in the interlayer site of smectites.

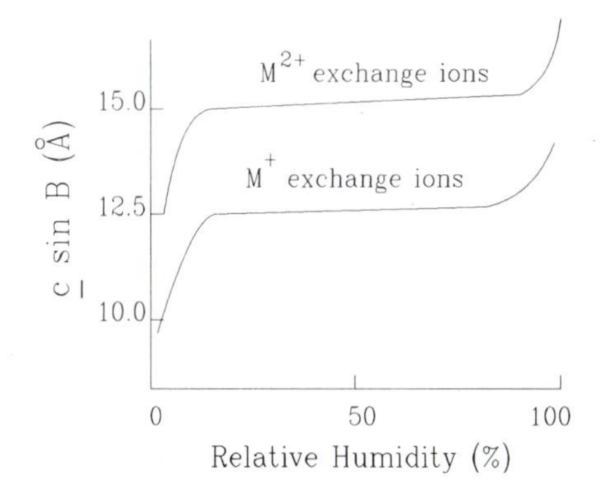

Fig. 3.25 Change in interlayer basal spacing of smectites as a function of relative humidity, showing the dependence on the type of exchange ion, mono- or divalent (M^+ or M^{++}).

smectites is ethylene glycol ($HOCH_2CH_2OH$) which gives an overall interlayer repeat distance of 16.9 Å. This exchange activity is basically chemical but it can have enormous physical consequences.

In systems where the only vapour present is water, the retention of water molecules is a function of the relative humidity in the air and the ambient temperature. The higher the temperature or the lower the relative humidity, the less water will be held on the cation in the interlayer site (Fig. 3.24). The mono-valent cations tend to release water more easily than do divalent cations in the interlayer sites. This is seen as a function of relative humidity in Fig. 3.25.

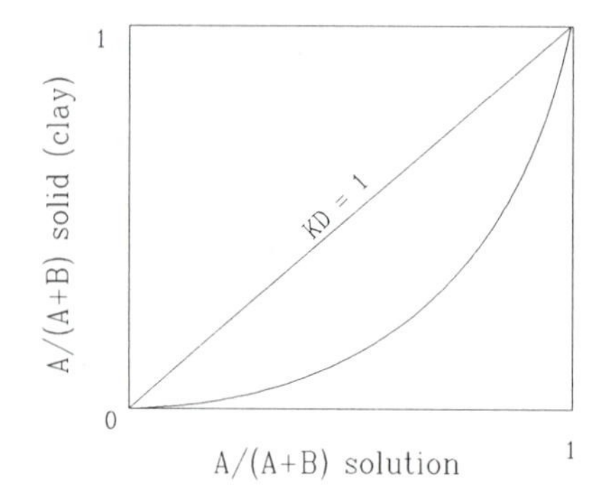

Fig. 3.26 Cation exchange partitioning coefficients. Distribution coefficient KD = 1 indicates no selectivity between solution and clay. The other curve indicates a preference for A in the solution. Hence, the clay tends to absorb B preferentially. For example, at near half of A in solution (that is, solution A/(A + B) = 0.5) the clay present in the solution will have only 0.2 A fixed on it, the rest of the ions on the clay are B (that is, clay A/(A + B) = 0.8).

Cation exchange selectivity

The interlayer ions are loosely held, so much so that they can be interchanged in aqueous solution which contains a high proportion of another ion. The exact ratio of ionic species in the solution will determine those in between the layers of the swelling clay. The proportion of one to another in each phase (aqueous and clay) determines the cation exchange isotherms or selectivity coefficients. The proportion of each ion in each site reflects the chemical activity of the ion in each substance. By thermodynamic definition, when the solution and the clay come into chemical equilibrium, the activity of the ions in each phase is the same. However, some ions prefer the clay to the solution when in the presence of other ions. For example, calcium much prefers clay over sodium or potassium. In general, the divalent ions are preferred in swelling clay interlayer sites to monovalent ones. The reverse is true for micas (with high charges). Figures 3.26, 3.27 and 3.28 show typical relations of cation preference in exchange between ions in aqueous solution and ions on a swelling clay.

It is important to realize that the diagram shows relations at equilibrium, when all portions of the system are at constant values. If one changes the composition of the solution, for example, the clay interlayer ion composition will also change. For example, if the initial solution is at x, with 50% cation A, the clay composition will be 20% A. If one changes the solution composition to y, 75% A then the clay will contain 50% A. Both will increase their A content but not in the same proportions.

The problem is slightly more complicated when other ions are added to

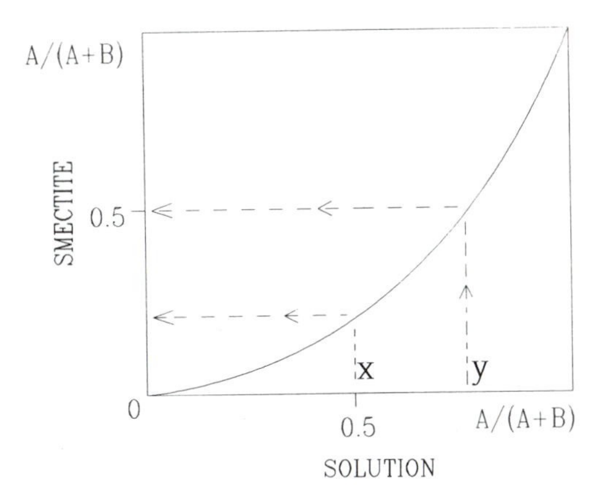

Fig. 3.27 The changing proportion of A found on clay depending upon the composition of the solution. Here B is favoured by the clay over A.

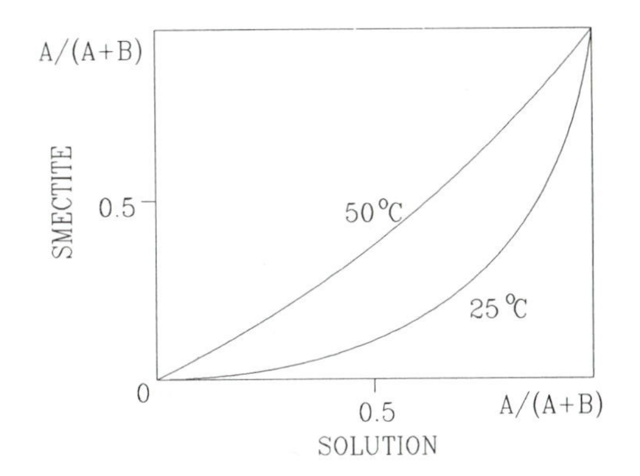

Fig. 3.28 Effect of temperature on the exchange coefficient. Here higher temperature tends to efface the selectivity of ion B by the clay.

the solutions but the results follow the same principles. If hydrogen ion content varies, it will compete with the other cations for places on the clay interlayer sites. Thus if one plots cation content against pH, the relations will change as a function of hydrogen ion content but not linearly if the cation content is not plotted in log units as in hydrogen content. This may seem evident but there are many studies in the clay literature which overlook this point.

The concept of ion exchange and cation partitioning is very important to studies which try to predict the chemical affinities and pathways during aqueous migrations.

The above outline of the mineral structures indicates the types of layer configuration, the 1:1, 2:1, 2:1 + 1 and the mica, 2:1 + interlayer types. In the 1:1 structures, there is a limited range of substitutions. Kaolinite is an almost pure Al-Si mineral. The trioctahedral equivalent shows a wide range of substitution types. In the 2:1 pyrophyllite structure, the mineral is equally almost purely Al-Si. Talc, the trioctahedral equivalent, shows divalent ion substitutions. The 14 Å, 2:1 + 1 chlorite structures show a wide range of substitutions. The widest range of substitutions are found in the different structures based upon the 2:1 + interlayer mica derived structure.

One can derive almost all of the minerals called clays from these models.

3.4 CLAY MINERAL GROUPS

The different mineral groups are identified by their structural arrangement (2:1, 1:1, etc.) and the elements found in the different polyhedral coordination layers. These chemical substitutions tend to modify the dimensions of the clay mineral structure. Some ions are effectively larger than others, that is, the outermost distance at which the coordination electrons occur defines the effective diameter of the ions. The change in structural cell dimension has limits, as one would expect. These limits define the boundaries between the clay mineral groups. They are the cause of the gaps in continuous compositional change called *solid solution*. These gaps are the result of dimensional incompatibility between coordination layers. The dimensional mismatch arises from the change in effective diameters of the ions or effects of attraction-repulsion of charged species aggravated by excessive substitutions of ions with different charges, that is, R^{3+} for R^{2+}. The general causes of these limits are, of course, outlined in most basic chemistry textbooks.

3.4.1 Dioctahedral minerals

Mica-like minerals (2:1 + interlayer)

It is perhaps unusual to start with the clay group whose limits are the most difficult to define but there is an underlying necessity to do this if one considers the logic of the mineral structures. Micas, as explained above, are 2:1 + interlayer ion structures. The substitutions in the tetrahedral and/or octahedral layers give rise to an overall charge imbalance on the 2:1 unit which is satisfied by the existence of a cation in between these layers. A sufficiently high charge on the 2:1 layers (greater than 0.9) makes a chemical and structural closure of one layer with its neighbour, that is to say, no polar ions can be inserted between the layers to change the structural dimension and bulk composition of the clay. Thus the basic repeat distance is near the 10 Å value of the 2:1 unit layer of two tetrahedra coordinated to one octahedral layer.

The clay mineral, mica-like phases are quite similar to high-temperature true micas, but not quite so. The clay micas or mica-like minerals have a charge on the 2:1 unit of less than 1.0. The second difference is that they are often intimately interrelated with smectite 2:1 layers. This intimate association, called **interlayering** or **mixed layering**, mixes different charged 2:1 units in the same crystal structure. This tends to confuse determination of the chemical limits of the pure mica-like phase because it is difficult to detect small quantities of the smectite in the presence of large quantities of mica-like layers. Thus, chemical determinations of the composition of these mica-like phases often include mixed-phase assemblages and the rigorous chemical limits established for other silicate phases, such as garnets or amphiboles, are much more difficult to find. Therefore, the estimation of the limits and range of composition given in the following section on micas is highly interpretative. This means that if one looks in the clay mineral literature for the composition of a specific phase, it might well fall outside of the limits set here. However, in most cases this will be due to the inclusion of multiphase (interlayer) material in the determination of the assumed one-phase sample.

Illite – glauconite, celadonite

All of the soil and clay mineral equivalents of micas are uniquely dioctahedral. The unhydrated, firmly fixed interlayer ion is always predominantly if not exclusively potassium. The relatively large size of the potassium ion props open the 2:1 layers to a 10 Å (001) basal spacing instead of the 9.6 Å of talc and pyrophyllite or the sodic and calcic micas paragonite and clintonite. This peculiarity is one to be remembered for purposes of mineral identification.

Sodic and calcic micas are not stable at low temperatures, nor are the trioctahedral micas or minerals with a high interlayer ion occupancy (1.0 per unit cell).

The low-temperature clay mica-like mineral groups are illites and celadonite-glauconites. The major difference in the two types of micaceous clay minerals is the site of substitution in the 2:1 structure which gives rise to the interlayer ion site occupancy. The illites are close to the muscovite structure and composition, $KAl_2(Si_3,Al)O_{10}(OH)_2$. The site of charge imbalance for the 2:1 layer unit lies in the substitution of slightly less than one Al for an Si ion in the tetrahedrally coordinated site. Illite is thus aluminous as little charge-inducing substitution occurs in the octahedral site. There is a certain amount of Mg and Fe^{2+} in the illite compositions

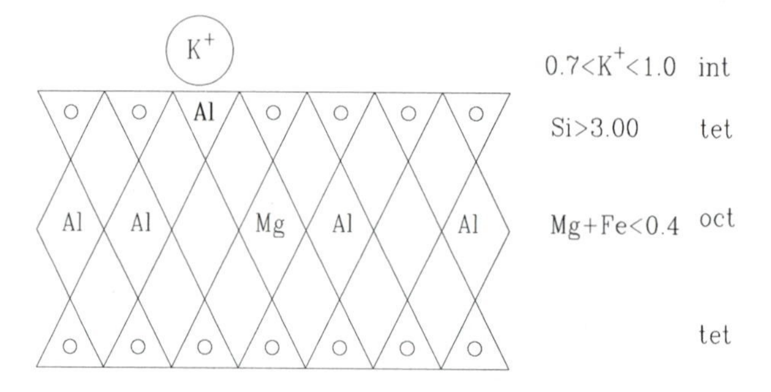

Fig. 3.29 Compositional relations and substitutions in illite, a low-charge, dioctahedral mica-like mineral.

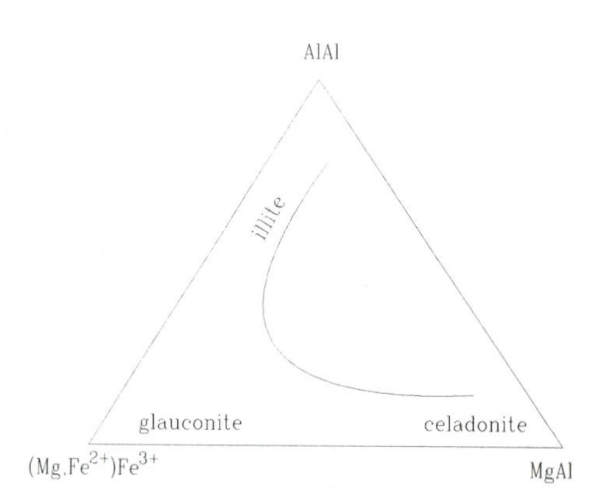

Fig. 3.30 Compositional relations of the low-temperature, mica-like minerals (all potassic) illite, glauconite and celadonite. The elements found in the octahedral site are given to distinguish the mineral types. Potassic, micaceous clays are found only to the left of the curved line.

due to R^{2+} = Al substitution. This substitution can contribute to in part to the interlayer charge imbalance through the $R^{2+} = Al^{3+}$ octahedral site substitution (Fig. 3.29). Due to a lower interlayer charge, illites are more siliceous than the high-temperature micas muscovite-phengite.

The celadonite-glauconite minerals are dominated by substitution in the octahedrally coordinated site which gives rise to interlayer charge. The minerals have a strong Mg, Fe component. Further, the substitution of Fe^{3+} for Al^{3+} is quite common in these minerals. The combination of di-and trivalent iron produces a typical green color in the glauconite-celadonites.

In a very general way, one can represent the three mineral family types on a triangular compositional plot (Fig. 3.30). The end member vari-

Table 3.3 Composition of mica-like phases (cations per $O_{10}(OH)_2$)

Illite *Typical*		*Range*	*Glauconite* *Typical*		*Range*	*Celadonite* *Typical*		*Range*
Interlayer sites								
K	0.77	0.6 −0.80	K	0.8	0.75–0.88	K	0.84	0.61–0.92
Na	0.01	0 −0.07	Na	0	0.01–0.07	Na	0	0 −0.13
C	0.02	0 −0.06	Ca	0	0 −0.07	Ca	0.03	0 −0.12
Octahedral sites								
Al	1.63	1.22–1.77	Al	0.4	0.36–0.50	Al	0.78	0.07–1.22
Fe^{3+}	0.03	0.03–0.45	Fe^{3+}	1.0	0.35–1.55	Fe^{3+}	0.47	0.36–1.14
Fe^{2+}	0.05	0 −0.22	Fe^{2+}	0.4	0.04–0.51	Fe^{2+}	0.21	0.12–0.26
Mg	0.30	0.15–0.36	Mg	0.2	0.10–0.51	Mg	0.68	0.48–1.04
Ti	0.04	0 −0.06	Ti	0	0 −0.01	Ti	0	0 −0.02
Tetrahedral sites								
Si	3.40	3.18–3.70	Si	3.90	3.54–3.93	Si	3.88	3.50–4.00
Al	0.60	0.30–0.82	Al	0.10	0.07–0.46	Al	0.12	0 −0.49
Cell dimensions								
(001)	10.1 Å			10.1Å			10.1Å	
(060)	1.50 Å		1.512–1.517Å			1.507–1.509Å		

ables in the figure are the dioctahedral ion pairs AlAl (illite), $MgFe^{3+}$ (glauconite) and MgAl (celadonite). Muscovite occurs at the AlAl pole but is not really a clay mineral because it does not form at low temperatures and has a higher potassium content, that is, 1.0 atoms per $O_{10}(OH)_2$ anicons, than do clays which have 0.7 to 0.95 atoms. Glauconite compositions are dominated by iron, very much of it being trivalent with some divalent ions. Celadonite is MgAl-rich. It is not certain whether there is a fully continuous series of compositions between illite and glauconite but there is certainly not one between celadonite and illite. It is probable that there is a continuous series of compositions between glauconite and celadonite.

The substitutions give rise to variations in the cell dimensions. The larger divalent iron ions (Fe^{2+} = 0.92 Å radius) increase the cell dimensions when they replace the aluminum ions (Al^{3+} = 0.39 Å). Magnesium has a lesser but important effect (Mg^{2+} = 0.72 Å). The most dramatic change in cell dimension is in the plane of the sheet structure (*a* and *b* crystallographic directions) and the dimension perpendicular to this is less affected ($c \sin \beta$ dimension).

Table 3.3 gives typical chemical analyses for the three minerals, the range in cell dimensions and the ranges in chemical substitutions which commonly occur.

Table 3.4 Clay mica polytypes

2M1 *hkl*	*d (Å)*	*I*	*1M* *hkl*	*d (Å)*	*I*	*1Md* *hkl*	*d (Å)*	*I*
002	10.0	>100	001	10.1	100	001	10.1	>100
004	5.02	50	002	5.03	37	002	4.98	50
110	4.48	50	020	4.49	90	020	4.48	90
$11\bar{1}$	4.46	60						
021	4.39	15						
111	4.30	20	$11\bar{1}$	4.35	30			
022	4.11	15	021	4.11	15			
112	3.97	12						
$11\bar{3}$	3.89	40						
023	3.74	30	$11\bar{2}$	3.66	60			
$11\bar{4}$	3.50	45						
006}			003}			003	3.33	90
024}	3.35	>100	022}	3.36	100			
114	3.21	50						
025	3.00	50	112	3.07	50			
115	2.87	35	$11\bar{3}$	2.93	6			
$11\bar{6}$	2.80	20						
131	2.59	50	023	2.68	15			
202	2.56	90	130	2.58	50	130	2.57	90
008	2.51	20	131	2.56	90	113	2.46	30

CEC (cation exchange capacity) The cation exchange capacity (CEC) of the mica-like minerals is low, between 5 and 10 meq of charge per 100 g of clay. Such an exchange capacity represents essentially absorption phenomena on the surface of the grains. This capacity is then determined largely by the grain size of the sample.

One must be careful with analyses in the literature for they often are made on minerals which are not pure mica-like phases, that is to say, a certain amount of other mineral layers are present. These are the interlayer minerals, which will be treated in Section 3.4.7. The intimate admixture of smectite as an interlayered component gives lower potassium contents and higher silica contents than those given in Table 3.3.

X-ray diffraction All of the mica-like phases have similar basal spacings (near 10 Å). However, there are several different structural arrangements (polytypes) which are important in identification of the different minerals and hence their origins.

There are three polytypes found in the clay micas. They are all based upon a monoclinic cell. One is a two-layered polytype, called 2M1, another is a regular one-layer polytype, 1 M and the third form is a disordered, one-layer polytype, 1 Md. Table 3.4 gives the different diffrac-

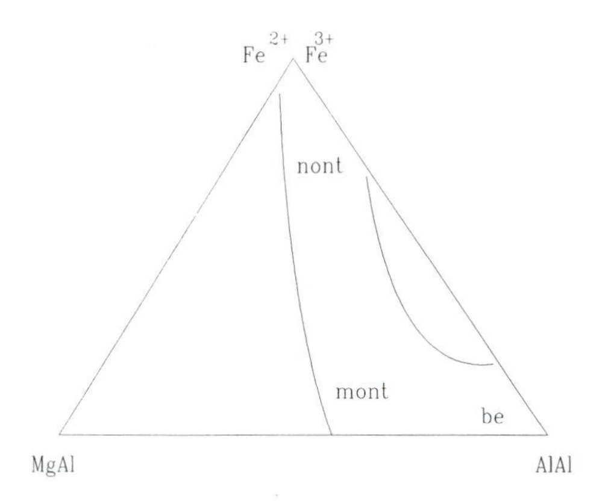

Fig. 3.31 Compositional relations of the dioctahedral smectites. Ions present in the octahedral site are used to distinguish between montmorillonite (mont), beidellite (be) and nontronite (nont) which are found between the curved lines.

tion maxima for the different structures. The celadonites and glauconites have a 1M polytype when they are pure mica composition. The illite has a 1Md or 1M polytype when it is of low-temperature origin and mixed with a minor smectite component and a 2M form when it is of a high-temperature origin.

Interlayering of mica layers with smectite gives 1M or most often 1Md polytypes to the structures.

Dioctahedral swelling clays (smectites)

The dioctahedral smectites are the phases which are often intimately associated (interlayering) with the mica-like minerals. As was the case for the mica-like minerals illite, glauconite and celadonite, the chemical analyses of the dioctahedral smectites which can be found in the clay mineral literature are often in fact those of a mixed-layer or interlayer phase, mica and smectite. The values given here are thus again rather strict, having been selected on the basis of rigorous X-ray diffraction determinations.

Composition

The minerals in the dioctahedral smectite group are distinguished mainly by their Al or reciprocally their Fe^{3+} content. Two main mineral groups are the nontronites (Fe^{3+}-rich) and the montmorillonite-beidellites (Al-rich). There is considerable, and possibly complete, compositional substitution between the nontronites and the beidellites. Also there seems to be significant but incomplete substitution between the montmorillonites and beidellites (see Fig. 3.31). Substitutions giving rise to charge on the

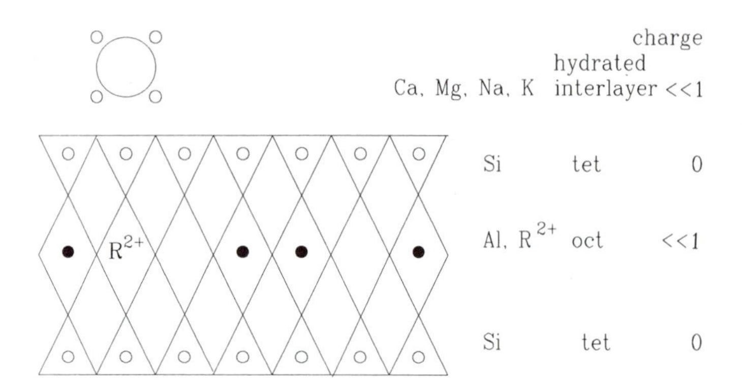

Fig. 3.32 Composition and substitutions in montmorillite (dioctahedral smectite). The origin of the interlayer charge (divalent R^{2+} for Al^{3+}) is found in octahedral substitutions. Tetrahedral composition remains silicic.

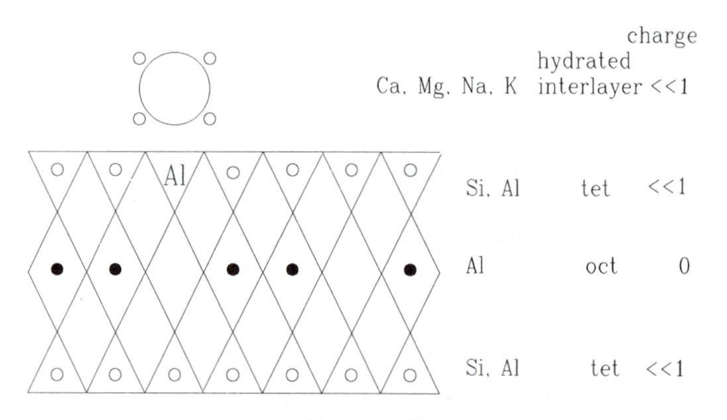

Fig. 3.33 Beidellite smectite Al^{3+} for Si^{4+} substitutions which create a layer charge imbalance (interlayer charge) originating in the tetrahedral layer.

2:1 layers and thus creating an interlayer ion occupancy, are derived in the tetrahedral site in the beidellite structure as in the illites. Charge substitution originates in the octahedral site in the montmorillonites as in the celadonites. The nontronites appear to have substitutions in both sites giving rise to charge on the 2:1 layer (Figs. 3.32–3.34).

The range of charge on the dioctahedral smectites seems to be rather large and in fact the subject of occasional controversy. However, it seems that the range can easily extend from 0.2 to 0.6 per 2:1 unit of $O_{10}(OH)_2$ anions. This being the case, the range in physical conditions which give rise to these phases might well be quite large, determining the range in chemical composition. These relations are not well defined at present.

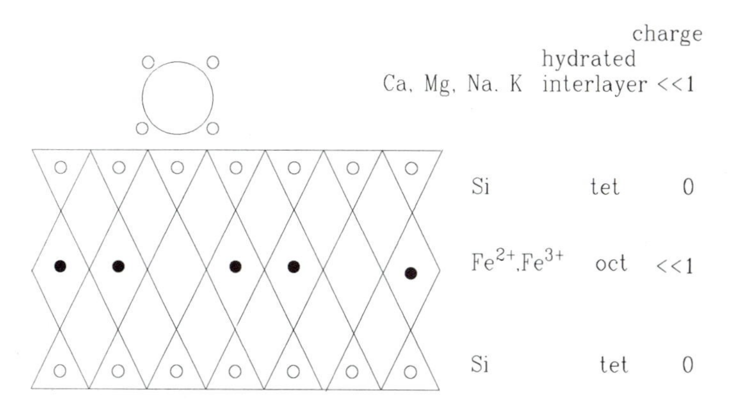

Fig. 3.34 Nontronite substitutions: Fe ions are dominant in the octahedral site. Charge imbalance is created by substitution of Fe^{2+} for Fe^{3+}.

Interlayer site
Whatever the interlayer charge on the structure, the cations which are present to compensate for the charge imbalance on the 2:1 layer are in all cases hydrated, that is, surrounded by water molecules at atmospheric conditions. This distinguishes the smectites from the mica-like minerals, where no water is present. This hydration gives a greater dimension to the 2:1 mineral layer in the $c \sin \beta$ direction. This dimension is readily changed by variations in the vapour pressure of water or by variations in temperature. The interlayer cations can have either two layers or one layer of water molecules surrounding them. At high temperatures, above 200°C, the interlayer ions are almost anhydrous. The hydration states of different cations are a function of the chemistry of the ion, that is to say, the power of attracting water molecules depends upon the electronic configuration of the ion. In general, the divalent ions remain hydrated to a larger extent than the monovalent ions common in natural clay environments (Na, K, Mg and Ca).

The most important feature of the smectite mineral family, as has been stated, is the capacity to accept and exchange hydrated cations and other polar molecules within the interlayer position. The type of molecule present is determined by the chemical activity of the molecule in the environment around the smectite mineral. This means that there can be a free and rapid exchange of material in the interlayer position of the smecitite minerals.

Table 3.5 shows the typical range of compositions for the three dioctahedral smectite types.

In natural smectite minerals the interlayer ion site is completely dependent upon the local and temporal chemistry of any aqueous solution

Table 3.5 Dioctahedral smectite minerals (cations per O_{10} $(OH)_2$)

Beidellite			*Montmorillonite*			*Nontronite*		
Typical		*Range*	*Typical*		*Range*	*Typical*		*Range*
Interlayer ion site								
M^+	0.46	0.45–0.47	M^+	0.27	0.21–0.62	M^+	0.26	0.23–0.74
Octahedral site								
Al	1.96	1.39–1.99	Al	1.56	1.17–1.57	Al	0.60	0.02–0.75
Fe^{3+}	0.02	0.02–0.49	Fe^{3+}	0.15	0.01–0.22	Fe^{3+}	1.06	0.17–1.45
Mg	0	0 –0.07	Mg	0.32	0.18–0.42	Mg	0.39	0.25–0.81
Fe^{2+}	0	0 –0.01	Fe^{2+}	0.01		Fe^{2+}	0.00	0.02–0.06
Tetrahedral site								
Si	3.48	3.46–3.58	Si	3.93	3.80–4.00	Si	3.98	3.41–4.00
Al	0.52	0.41–0.54	Al	0.07	0.00–0.20	Al	0.02	0.00–0.59
Cell dimensions								
(001) basal spacings are essentially the same								
b	8.99 Å			9.0–9.15 Å			8.94–9.08 Å	

with which the smectite clay is or was in contact. Therefore the species which are found in it in natural clays represent the local and most recent aqueous chemistry of the clay environment. This does not necessarily reflect the conditions which formed the smectite clay nor does the interlayer ion chemistry reflect the basic properties of the 2:1 clay structure. The physical properties of the clay can be changed in changing the interlayer ion species. As a result of this observation no attempt is made here to indicate in detail the species in the interlayer ion site in natural clays. M^+ is used to represent the total of ionic charges present as cations in the interlayer site.

3.4.2 Trioctahedral minerals

Talc

This mineral has only recently been recognized as a common constituent of the clay mineral assemblages. Previously, it was found in metamorphic or hydrothermal environments where it tends to have a large grain size, that is, greater than 2 μm. However, talcs form commonly in soils (Chapter 4) where they can have a complete range in Mg–Fe substitution but apparently little trivalent ion substitution occurs. Therefore the tetrahedral layer is uniquely composed of $4Si^{4+}$ units. The octahedral layer shows the full range of substitutions between 3Mg and $3Fe^{2+}$ (Fig. 3.35).

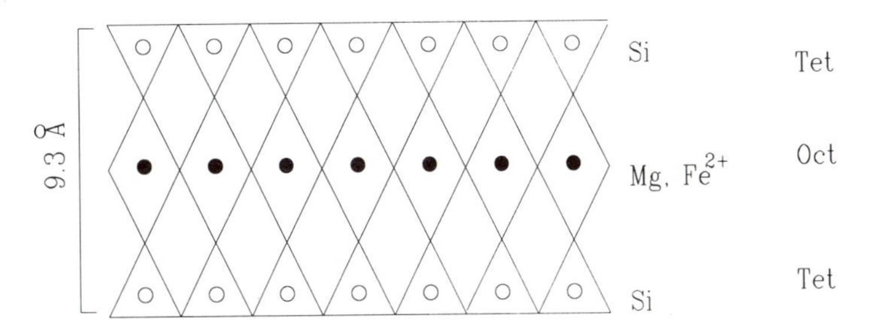

Fig. 3.35 Talc has an uncharged (electrostatically neutral) 2:1 trioctahedral structure.

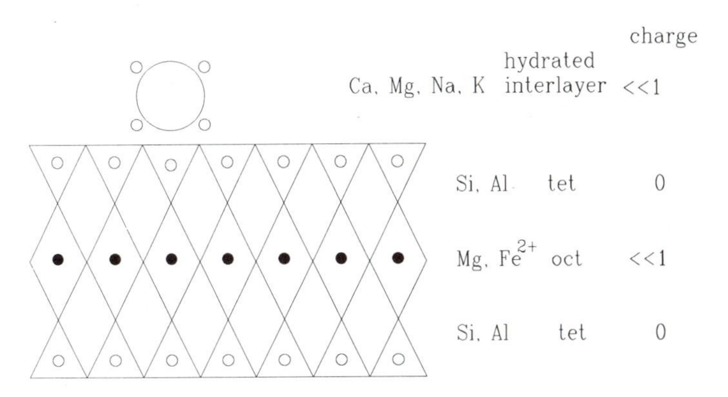

Fig. 3.36 Substitutions in stevensites (trioctahedral smectite). Either Mg or Fe^{2+} is normally found in the octahedrally coordinated site. Charge imbalance on the 2:1 structure comes from underfilling in the octahedral site.

Swelling trioctahedral 2:1 minerals

There are not enough data available at present to define clearly the limits of substitution in this family of minerals. Two main groups can be established with some confidence, hectorite-stevensite and saponites. The difference between the two mineral groups is in alumina content, hectorites being more aluminous. The hectorites are somewhat rarer than the stevensites.

Stevensite

The stevensites are predominantly magnesian smectites of rather low charge. In fact there is some doubt as to their true structure as smectitic minerals. The major substitution in the 2:1 structure giving rise to an interlayer charge is that of an octahedral vacancy type (Fig. 3.36). The

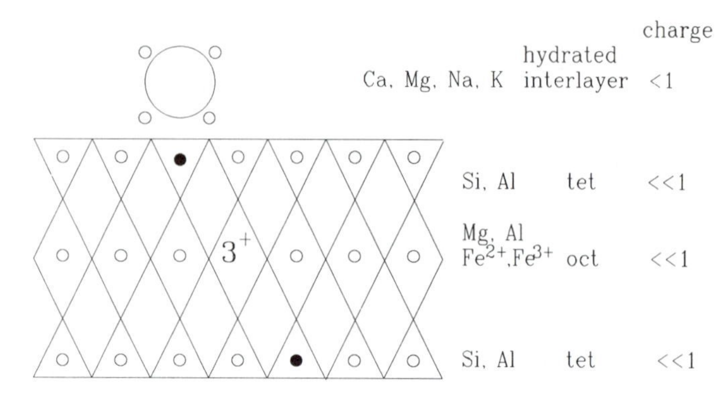

Fig. 3.37 Saponite (aluminous trioctahedral smectite) structure. Charge imbalance comes from substitutions in either the tetrahedral or octahedral site.

stevensite minerals are almost devoid of aluminium and Fe^{3+}. The substitution developing interlayer charge is of the type:

$$3Mg_4Si = 2.5Mg_4SiM^+$$

This substitution is also known in high-temperature micas. It is well known in synthetic micas but much rarer in natural minerals. It is exclusively found in magnesian minerals. Thus stevensites are almost exclusively composed of magnesium and silicon.

Al-saponite

Saponites are aluminous and of variable interlayer charge, following substitutions in both tetrahedral and octahedral sites (Fig. 3.37). Iron-bearing (ferrous) forms appear to be rapidly oxidized on contact with air and therefore estimation of ferric and ferrous iron content is difficult. It is assumed that the iron is almost exclusively ferrous in the stable structure.

Vermiculite

The mineralogy of this group is very poorly known in clay mineral environments. Most mineralogical studies have been performed on materials which are the result of hydrothermal alteration of biotites. They have grain sizes greater than 2 μm and a high-temperature origin. Vermiculite chemistry is closely related to the precursor biotites. These minerals were formed in a very special chemical and physical environment which has little to do with the minerals which form in soils which form the bulk of vermiculites.

The soil clay vermiculites are certainly of a wide variety of compositions and structures. Much material designated as vermiculite is of smectite

Table 3.6 Trioctahedral clays

Talc★		*Saponite*	*Stevensite*	*Soil vermiculite*
Interlayer charge				
0		0.42	0.16	0.6–0.9
Octahedral				
Mg	0.0–3.0	2.86	2.89	0.6–1.0
Fe^{2+}	0.0–3.0	0.0	0.02	0.3–1.0
Al^{3+}	0.0	0.08	0.0	0.4–1.0
Fe^{3+}	0.0	0.01	0.01	0.1–0.6
Ti	0.0	0.0	0.0	<0.1
Tetrahedral				
Si	4.0	3.75	4.00	2.9–3.3
Al^{3+}		0.25	0.0	1.1–0.7

★These values are valid for low temperature minerals only. Metamorphic talcs can contain several percent Al_2O_3 but they are not ferrous.

origin. The smectite has been altered by the insertion of a poorly expanding interlayer component, decreasing its expandability and exchange capacity. Often aluminium hydroxide and magnesium hydroxide units are interlayered in an irregular manner between the 2:1 layers.

There are both di- and trioctahedral vermiculites, of largely unknown composition. In fact the exact definition of vermiculite is hard to ascertain. A soil clay mineral with a basal spacing slightly below 15 Å under humid conditions which shows an increase towards 15.5 Å upon glycollation is often considered to be a vermiculite. Heating vermiculites releases water, but with much more difficulty than in smectites. Often heating to 300–400°C is not sufficient to render the interlayer ions anhydrous. The typical clay vermiculite is difficult to expand and difficult to dehydrate. Normally the charge on the structure is high, above 0.6 and below 0.9 for $O_{10}(OH)_2$ anions. Thus one can describe a vermiculite as being a high-charge smectite which expands less than a smectite. The exact structural and chemical configurations are not at present sufficiently defined for most clay vermiculite minerals given here. Representative chemical compositions of these trioctahedral minerals are given in Table 3.6.

3.4.3 1:1 dioctahedral clays (kaolinite, dickite, nacrite halloysite)

The kaolinite structure is important in that it only represents the kaolinite family of minerals. No other clay mineral types can be derived from it.

There is little substitution in the structure, and the kaolinite minerals are chemically 'pure', giving them useful properties in industrial use.

The 1:1 (tetrahedral–octahedral layer) structure has a tetrahedral layer composed uniquely of Si ions (see Fig. 3.21). This layer, as in all clay structures, is linked to an octahedrally coordinated layer of cations through a common layer of shared oxygen ions. Only two of the three sites in this layer are occupied by, again, almost uniquely aluminium ions. Small amounts (ca. 2%) of the sites can contain ferric iron. Thus the cations show an almost constant composition, with no substitutions. The apical anions of the octahedral layer are hydroxyls. These 'outer' hydroxyls have a slightly different reactivity that those hydroxyls in the centre of the octahedrally-coordinated layer.

The mineral formula is $Al_2Si_2O_5(OH)_4$; the thickness of the two layers of coordinated cations is 7 Å. The tetrahedral layer is near 3 Å thick and the octahedral layer near 4 Å. The typical 'basal' spacing defines this 7 Å mineral. Its cell dimensions are:

$$c \sin \beta \ (001) = 7.16\,\text{Å}$$
$$b = 8.93\,\text{Å}$$

These minerals are of one almost constant composition concerning the cations, Si in the tetrahedral layer and Al in the octahedral layer. The Fe_2O_3 is found in poorly crystalline kaolinites, up to 2%, but this minor substitution seems to be reduced as the minerals are better crystallized.

The names dickite and nacrite indicate a different stacking structure (polytypes) of the basic 7 Å kaolinite mineral.

Halloysite, however, is a kaolinite with an extra layer of water molecules between the 7 Å layer sheets which change its basal spacing (c sin β, or (001)) to near 10 Å. The water is easily removed near 100°C and in some cases the dehydration is irreversible. The halloysites also appear to contain high amounts of Fe_2O_3, up to 15% This distinguishes them from the kaolinites. Halloysites also have a tubular structure of rolled sheets instead of the typical flat sheet structure of kaolinites and most phyllosilicates.

The different structural states, that is, stacking patterns which give polytypes, of kaolinite allow one to identify the different subspecies of the kaolinite family. All are monoclinic. Table 3.7 shows the diffraction maxima for the various types of kaolin minerals.

3.4.4 1:1 trioctahedral minerals

The type structure is given as that of serpentine, a magnesium and silicon phase. There are five major ions found in the other minerals with the same

Table 3.7 Kaolin minerals

Kaolinite *1 layer, 1 M* *hkl*	*d* (Å)	*I*	*Dickite* *2 layer, 2 M* *hkl*	*d* (Å)	*I*	*Nacrite* *6 layer* *hkl*	*d* (Å)	*I*	*Halloysite* *hkl*	*d* (Å)	*I*
001	7.16	>100	002	7.15	100	002	7.18	100	001	10.1	10
020	4.46	40	020			111	4.44	30	02	4.46	8
$1\bar{1}0$	4.36	50	110	4.44	30	202 }	4.36	80			
$1\bar{1}1$	4.18	50	$11\bar{1}$	4.36	30	110 }					
$1\bar{1}\bar{1}$	4.13	30	021	4.26	20	$11\bar{2}$ }	4.13	70			
$02\bar{1}$	3.85	40	$11\bar{2}$	4.12	65	200 }					
021	3.74	20	111	3.95	10	111	3.94	20			
002	3.57	>100	022	3.79	60	004	3.59	80			
$1\bar{1}1$	3.37	40	$11\bar{3}$ }	3.58	90	204	3.48	20	003	3.40	5
$11\bar{2}$	3.14	30	004 }			112	3.41	20			
$1\bar{1}\bar{2}$	3.10	30	112	3.43	20	202	3.06	20			
$1\bar{3}0$ }	2.56	60	$11\bar{4}$	3.10	8	113	2.93	10			
201 }			113	2.94	8	020 }	2.57	10	20	2.56	5
131	2.53	40	024	2.79	10	$31\bar{2}$ }					
$1\bar{3}\bar{1}$			13			306					
200	2.49	80	$20\bar{2}$	2.56	35	multi	2.52	B			
003	2.37	60	114	2.50	50	022 }	2.43	60	04	2.37	3
			006	2.38	15	$31\bar{4}$ }				2.23	3
060	1.489	40	060	1.489	40	060	1.489	40			

structure: Fe^{2+}; Mg; Al; Fe^{3+}; and Si. The substitutions in the octahedral and tetrahedral sites serve to distinguish the clay mineral groups found in nature. More complete substitutional series have been found to occur in synthetic minerals formed in the laboratory. All substitutions give a neutral lattice structure. Hence there are no expanding or swelling 1:1 clay minerals.

Table 3.8 gives the major mineral types as a function of their ionic content and Fig. 3.38 shows the substitutional range of the minerals found in nature.

Substitutions

In the trioctahedral 1:1 structures, with a 7 Å repeat distance there is a complete range of Fe^{2+}–Mg substitution. There is a more limited range of Al^{tet}–Al^{oct} substitution, between slightly aluminous serpentines, about 10% Al in the cation sites to near amesite composition, $R^{2+}Al_2(SiAl)O_5(OH)_4$. The low-temperature clay minerals tend to have an aluminium content in the middle of this range. Scarcity of analyses does not permit one to give a table with formulae and variations of elemental concentrations in the various sites. Table 3.9 gives some typical analyses of each mineral type.

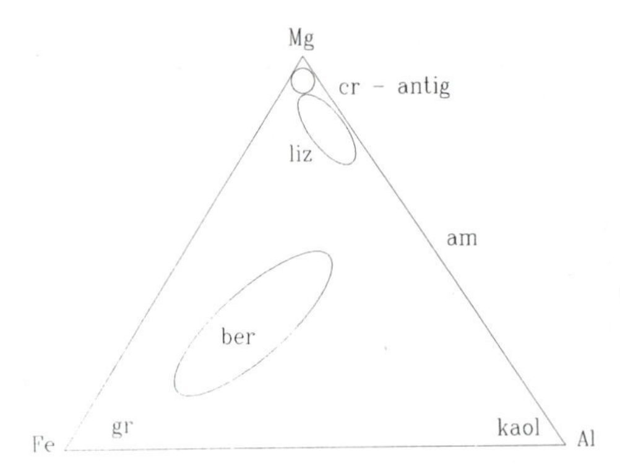

Fig. 3.38 Chemical relations between different low-temperature 1:1 7 Å trioctahedral minerals. liz = lizardite, cr = chrysotile, antig = antigorite, am = amesite, ber = berthiérine, gr = greenalite, kaol = kaolinite. See Table 3.8 for a mineralogic definition of these minerals. Kaolinite, berthiérine and greenalite are the only minerals to occur under conditions of clay mineral genesis (that is below 200°C).

Table 3.8 7 Å trioctahedral minerals

$R^{3+}R^{2+}$		*b (Å)*	*(001) (Å)*	*Environment*
Mg –	chrysotile antigorite	9.25	7.32	metamorphism of ultra basic and basic rocks
Mg ≫ Al	lizardite	9.21	7.14	metamorphism of basic and ultrabasic rocks
Mg = Al	amesite	9.21	7.02	metamorphism of basic and ultrabasic rocks
Fe^{2+}	greenalite	9.70	7.21	low grade metamorphism of iron ores
Fe^{2+} Al	berthiérine	9.38	7.11	sediment–sea water interface
Fe^{2+} Fe^{3+}		9.10	7.06	
Fe^{2+} Al Mg	berthiérine	"	"	diagenetic formation in buried sediments

It can be seen in Fig. 3.40 that the major iron-bearing minerals are of low-temperature origin and have a reasonably large range of composition. The aluminous 1:1 minerals, kaolinites, have very restricted compositional ranges. They are formed under low- and high-temperature conditions. The low-temperature phases which contain Fe, Al and Mg in significant quantities should all be called berthiérines according to the nomenclature in present use. They are formed in two environments, in pelletal replacements in sediments at the sea-water interface or under conditions of burial

Table 3.9 Typical mineral compositions

	Chrysotile	*Lizardite*	*Amesite*	*Greenalite*	*Berthiérine*
Octahedral					
Mg	5.88	5.82	3.28	0.61	0.27
Al(oct)	0.06	0	2.00	0.12	1.84
Fe^{2+}	0.01	0.01	0.67	4.26	3.32
Fe^{3+}	0	0.13	0	0	0.37
Tetrahedral					
Si	4.00	3.91	2.01	4.00	2.20
Al	0	0.08	1.99	0	1.80
Fe^{3+}	0	0.01	0	0	0

diagenesis (1–5 km depth) where they are a result of mineral reaction and recrystallization of the sediments. In sedimentary rocks, the berthiérines are eventually transformed into 14 Å 2:1 + 1 structures and called **chlorites**.

The XRD characteristics for trioctahedral 1:1 minerals are given in Table 3.10. These are the common features of diagrams for the most common polytypes in the given compositional range. The three minerals are lizardite, of low alumina content, berthiérine of intermediate alumina content, and amesite which has a specific, fixed composition of alumina content. The lizardite and amesite are high-temperature minerals. The data for berthiérine are the only set applicable to 1:1 trioctahedral clay minerals.

One may remark in passing that even though it is not strictly a clay mineral in its origin, chrysotile is of interest because of its crystal morphology. Chrysotile is a major component of asbestos. This material has a fibre-like structure. Chrysotile has needle-shaped crystals. These needles irritate the internal respiratory organs of humans, and undoubtedly other animals, to the point of creating favourable conditions for the development of cancer and other malignant growths. Almost all other clays have a flat, plate-like shape which has not been demonstrated to harm internal organs of animals or humans.

3.4.5 2:1 + 1 minerals (14 Å chlorites)

The 14 Å chlorite structure is the best-known form of this mineral group. In sedimentary rocks it seems that the low-temperature phases are all of the 1:1 7 Å type. At depths which give temperatures near 100°C it appears that these 7 Å chlorites become 14 Å phases. However, one can find 2:1 + 1 14 Å chlorites as authigenic phases in soil profiles. Therefore it is not at

Table 3.10 XRD data for 1:1 aluminous trioctahedral minerals

Berthiérine IM (A)			*Amesite 2H2 (D)*			*Lizardite 1T (C)*		
hkl	*d (Å)*	*I*	*hkl*	*d (Å)*	*I*	*hkl*	*d (Å)*	*I*
001	7.15	100	002	7.0	100	001	7.22	60
020	4.65	40	020, 110	4.60	25	100	4.63	40
110	4.53	20	021, 111	4.38		101	3.87	30
002	3.58	85	022, 112	3.84	30	002	3.62	50
022	2.84	10	004	3.51	100	102	2.85	10
201, 130	2.68	40	023, 113	3.27	25	110	2.67	10
112	2.56	7	024, 114	2.79	10	111	2.51	100
201, 131	2.51	3	131, 201	2.60	40	003	2.42	10
040	2.34	5	132, 202	2.47	80	112	2.15	60
203, 132	2.02	10	025, 115	2.39	12	113	1.79	40
			006	2.34	20			
			133, 203	2.30	25			
			134, 204	2.11	25			

present certain whether or not there is a continuous stability of 2:1 + 1 14 Å chlorites from surface to metamorphic conditions or a sequence going from 14 Å to 7 Å to 14 Å as temperature increases during burial and diagenesis.

In the cases where chlorites are formed in soils, the compositions show a clear trend of di-trioctahedral substitutions. Chlorites in sedimentary rocks formed under conditions of diagenesis have the same compositional trends. In general, as temperatures increase towards metamorphism (above 200°C) the chlorite composition becomes uniquely trioctahedral. This change seems to be independent of the proportion of Fe compared to Mg in the minerals. The compositional relations are shown in Fig. 3.39 where the general compositional field of low-temperature and high-temperature chlorites found in sedimentary rocks is shown. The examples used to construct this diagram are chlorites found in pelitic sedimentary rocks, that is, they are found in the presence of aluminous minerals such as illite and mixed-layer illite/smectite minerals. Thus their chemical environment is aluminous.

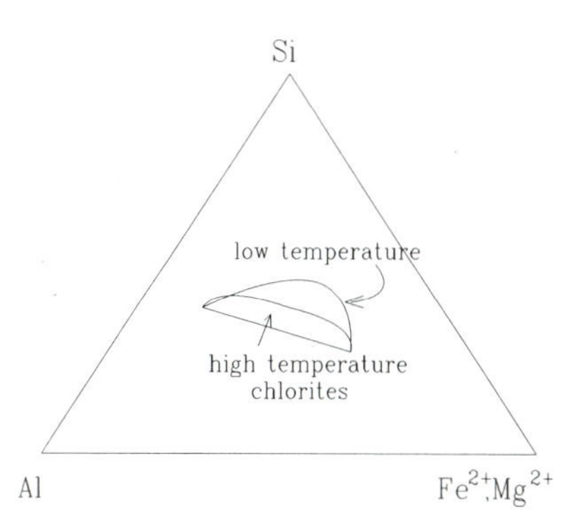

Fig. 3.39 Representation of low- and high–temperature chlorites in Si-Al-R^{2+} coordinates. An increase in temperature during diagenesis or low-grade metamorphism decreases the range of chlorite compositions.

Table 3.11 XRD for chlorite polytypes

Polytype		*Ia*	*IIb*			*Ib*
d (Å)	*hkl*	*Intensities*		*d (Å)*	*hkl*	*int*
2.66	20$\bar{1}$	44	29	2.67	200	75
2.65	200	116	6	2.62	201	41
2.59	20$\bar{2}$	79	98	2.50	202	316
2.55	201	36	148	2.33	203	50
2.44	20$\bar{3}$	26	134	2.13	204	210
2.39	202	228	105	1.95	205	49
2.26	204	64	94			
2.20	203	33	28			
2.07	205	52	44			
2.01	20$\bar{4}$	168	176			

In diagenesis, it appears that 14 Å 2:1 + 1 chlorites can have one of three polytypes whose presence is a function of temperature history of the sediment. Polytype IIb is the most stable form and is found in chlorites from metamorphic rocks. Lower-temperature chlorites are of type Ia or Ib. Table 3.11 shows XRD data for chlorite polytypes.

3.4.6 Sepiolite and palygorskite

These two minerals are closely associated in their mineral structure, chemical composition and geological occurrence. X-ray diffraction gives two distinct patterns for the two minerals (Table 3.12) but the funda-

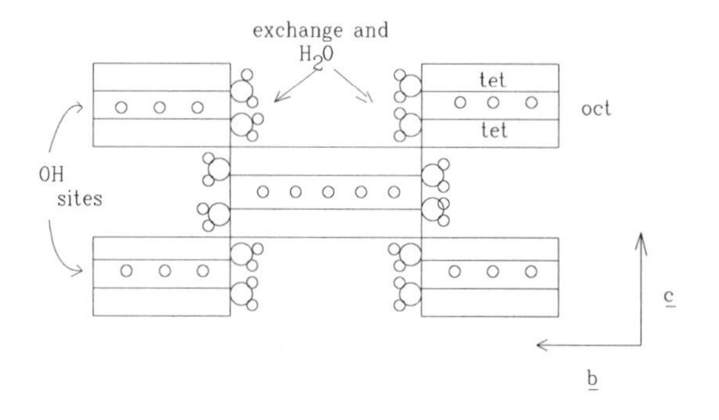

Fig. 3.40 Structure of sepiolite and palygorskite. 2:1 units are not sheets but ribbons (extended perpendicularly to the page) which are linked forming an open cavity, similar to that in zeolites, in which water and exchange ions can be accommodated. Crystalline water (OH units) is found in the octahedral layer of the 2:1 units. Interlayer exchange sites are found in the cavities.

Table 3.12 X-ray diffraction data for sepiolite and palygorskite

Sepiolite			*Palygorskite*		
hkl	*d* (Å)	*I*	*hkl*	*d* (Å)	*I*
011	12.1	100	011	10.4	100
031	7.5	7	002	6.4	13
002	6.7	45	031	5.4	9
051	5.04	3	040	4.5	20
060	4.49	25	121	4.3	22
131	4.29	35	013	4.2	2
033	4.02	7	122	3.67	15
062	3.73	25	051	3.44	2
142	3.50	5	132	3.35	7
080	3.34	45	004	3.18	12
133	3.18	15	123	3.10	16

mental structural units are similar. Both sepiolite and palygorskite are formed from shortened sheets of octahedral–tetrahedral linkages which are alternately inverted (see Fig. 3.40).

The small lateral extension (*b* dimension) of the sheets gives rise to a linear, needle-like morphology (*a* dimension). The alternation of the sheet orientations creates channels in the structure which are occupied by water and exchangeable ions. The exchange capacity of sepiolite and palygorskite is less that of than smectites (20–60 meq) but their significant capacity to absorb organic molecules has led to their use as an industrial and domestic absorbant.

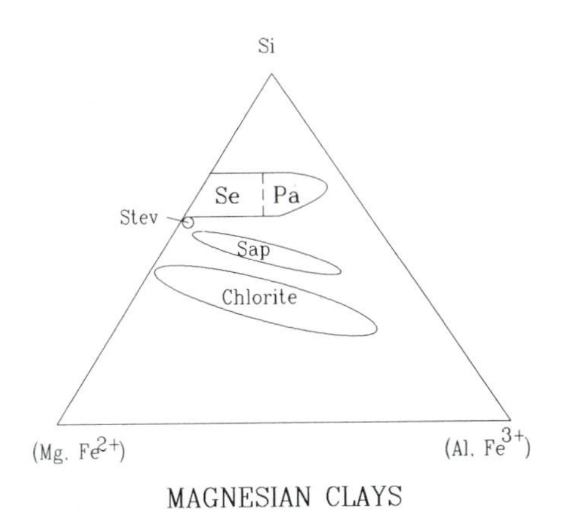

Fig. 3.41 Relationships of the chemistry for chlorite, sepiolite (Se), palygorskite (Pa) and saponite smectite clays as plotted in Si, divalent ion and trivalent ion coordinates.

Discussion continues concerning the exact structure of the two minerals. However, it seems clear that palygorskite has a consistently greater alumina content than sepiolite. It is probably basically a dioctahedral mineral, while sepiolite, which can be almost exclusively magnesian, is trioctahedral in basic character. The compositional relations of these two minerals are shown in Fig. 3.41, where they are compared to those of talc and saponite. The coordinates chosen are Si, R^{2+} (Mg and Fe^{2+}) and R^{3+} (Al and Fe^{3+}). Ferric iron is rare but when present it dominates the R^{3+} substitution and forms a mineral called **xylotile**.

The generalized mineral formulae are:

palygorskite	$(R^{2+}, R^{3+})_5(Si, R^{3+})_8O_{20}(OH_2)_4.\, M^+(H_2O)_4$
sepiolite	$(R^{2+}, R^{3+})_8(Si, R^{3+})_{12}O_{30}(OH_2)_4.\, M^+(H_2O)_8$

The division between sepiolite and palygorskite in their chemistry is made at Mg/Mg + Al = 0.6, sepiolite having a proportion greater than 0.6.

It is interesting to note that sepiolite-palygorskites are the most silica-rich of the ferro-magnesian clay minerals. All clays have roughly the same alumina substitutions but only sepiolite-palygorskites show a change in mineral structure in the sequence.

The exchange capacity is due to R^{3+} substitutions for Si^{4+} ions and possibly underfilling of the octahedral sites. All of the substitutional types appear in the minerals, Al = Si tetrahedral, $Al^{tet} = Al^{oct}$, Al + deficit octahedral, Fe^{2+} = Mg octahedral, $Fe^{3+} = Al^{3+}$ octahedral. The exchangeable ions can represent up to 60 milli-equivalents of cation exchange

Table 3.13 Mineralogical data for palygorskite and sepiolite

	Palygorskite		*Sepiolite*	
	Typical	*Range*	*Typical*	*Range*
tet				
Si	7.50	8.00–7.34	11.77	11.96–11.23
Al	0.50	0.00–0.66	0.23	0.04–0.46
oct				
Mg	1.78	1.29–2.81	6.98	7.96–6.98
Fe^{2+}	0.10	0.10–0.47	0.25	0.00–0.13
Al	1.62	0.95–2.35	0.02	0.00–1.37
Fe^{3+}	0.41	0.00–0.42	0.25	0.00–0.47
M^+	0.68	0.04–0.72	0.01	0.00–0.45
cell dimensions (Å)				
a =	5.2–5.24		5.24–5.30	
b =	18.0–17.8		26.77–27.2	
$c \sin \beta$ =	12.7–12.9		13.3–13.4	
CEC	20–60			

capacity (CEC). However, there is a small but significant amount of non-exchangeable Ca and K reported frequently which might be due to fixed, mica-like cation sites. Table 3.13 gives chemical compositional data for sepiolites and palygorskites.

The striking feature of the sepiolite-palygorskites is the intermediate nature in their chemical absorption properties. The minerals have intermediate exchange capacities, near 40 meq (between smectites and non-exchanging minerals). They also have several types of water present in their structure, crystalline OH, fixed H_2O molecules which are released at temperatures above 110°C and zeolitic water that is not chemically bound and is easily removed.

Figure 3.42 shows the temperatures at which these water species leave the structures of sepiolite and palygorskite.

3.4.7 Mixed-layer minerals

This is the most difficult and perhaps the most interesting of current clay subjects. The discovery of mixed-layer phenomena in the 1950s led to a new understanding of clay minerals and clay reactions. As indicated in the preceding chapters on mineral chemistry, each clay group has a certain range of chemical composition which can be found within each layer of the layer silicate structures. These are solid solution type substitutions which change the chemistry on a general scale throughout the crystal in

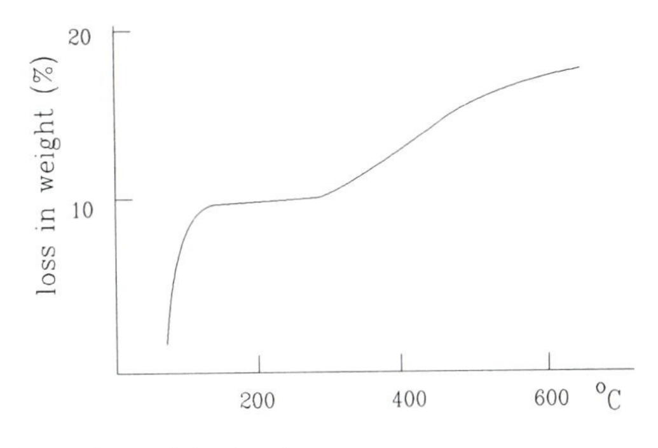

Fig. 3.42 Sepiolite-palygorskite heating curve.

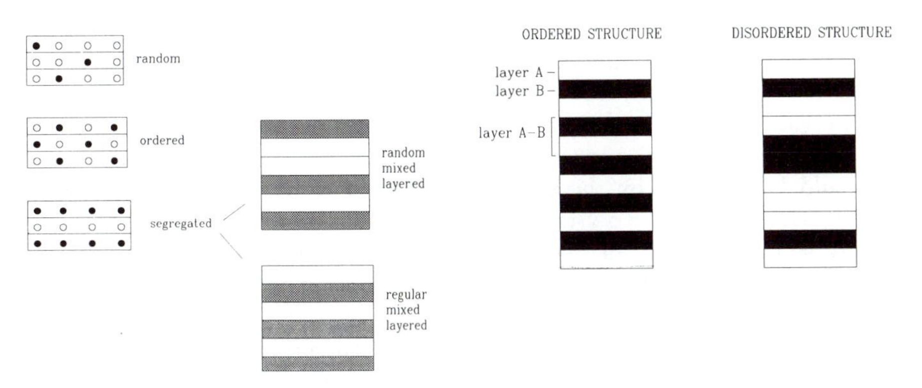

Fig. 3.43 Examples of different types of ordering of ions in the layers of sheet silicates. Segregations can occur in ordered arrays in all layers, or in different layers. Segregated ions in layers can form different layering types. Ordering of layering types can occur in a random manner or as ordered layers.

about the same proportions in each individual layer unit. There are generally no segregations of the different ions in the units of the structure. However, as suggested by the name 'mixed-layer mineral', it has been found that significant segregation can be found in some clay mineral crystals.

At present there is debate as to the exact location of ions and the composition of the layers in crystallites of mixed layer minerals, but there seems to be no doubt that some layers of the structures can have a composition decidedly different from other layers. Figure 3.43 gives an impression of the types of arrangement likely to arise in layer structures. One can imagine three basic types of arrangement of atoms in a regular layer structure: random distribution of the species (a); an ordering or

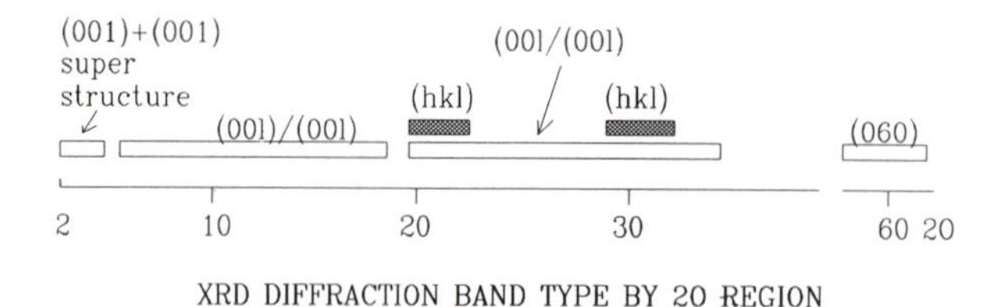

Fig. 3.44 Indication of the regions of an XRD pattern where the different types of diffraction maxima occur for mixed layer minerals.

regular alternation of the atoms in the sites in each layer (b); and a segregation of atoms by layer. This last will lead to mixed layering. The mixed layering, segregation by layers, can in turn be random or ordered according to the relations of the successions of the different types of layers in the crystal. It is important to remember that individual crystals in a clay sample can and will more than likely have slightly different compositions. If crystals of different composition are piled upon one another during the slide preparation process (see XRD section 2.1.2), a sort of mixed layering could be detected by the diffraction process which was not due to the variations within the crystals themselves. This latter possibility is not very probable in practice in that most clay particles are more than one layer thick and will not be artificially made into a large synthetic crystal on the diffraction preparation.

The interpretation of the diffraction patterns produced by mixed-layer clays is very difficult and the reader should consult a book dealing with this problem specifically. In general, the phenomenon gives rise to diffraction maxima which are found at positions intermediate between the poles of the constituents. Thus peaks of mica/smectite mixed-layer minerals (001 mica + 001 smectite) will be found between the basal spacings of the individual mica and smectite layers, 10 and 15 Å. A summary of the regions in diffraction angle for the different type of XRD peaks is given in Figure 3.44. The exact position of this combined band depends upon the proportion of each component, hence a means of determining the average composition of the mineral, and the state of ordering of the sequence of layers. There are certain rules of proximity which allow diffraction maxima to be found for various combinations of peaks. For example, if the distances between lattice maxima are not too great one can find a combination peak of 001/001 as well as a 002/001 band in the same pattern, and so forth. It is also possible to find a combined peak representing the sum of the two components. This occurs in samples where the two interlayered elements are highly ordered, that is, one layer is always followed by the other in the stacking sequence. This ordering gives a superlattice peak, and multiples thereof. For example, one can find a 25 Å

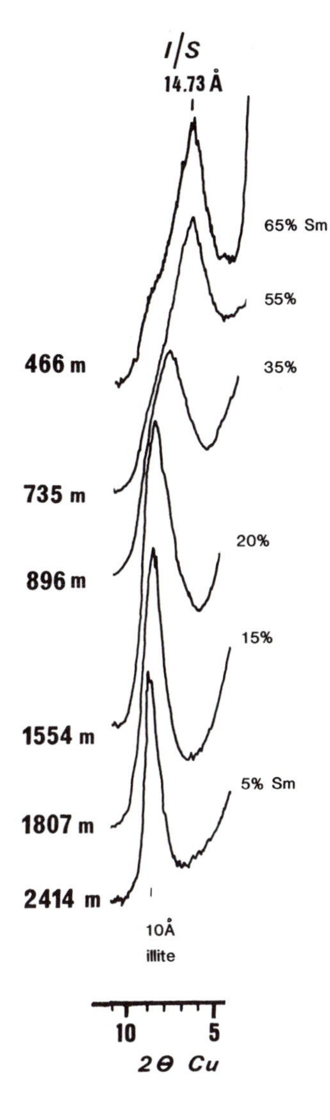

Fig. 3.45 Series of I/S mineral diffractograms in the 10–14 Å region for a deep well in a sedimentary basin. The depth at which the sample was taken is indicated at the left of the figure. The percentage of smectite in the illite/smectite mixed-layered phase present in the rocks is indicated on the right of the figure. The samples were run on the XRD diffractometer in the air-dried state giving a smectite interlayer spacing of 15.2 Å and an illite spacing of 10 Å. One can see the steady change in average peak position of the I/S minerals from 14.73 Å to 10 Å as the smectite content decreases with burial depth (increasing temperature and age of the sediment).

first-order peak for a perfectly ordered mica/smectite mineral in a hydrated state. Also one finds 12.5, 8.3, 6.3, etc., peaks which are (001) reflections of the combined 10 + 15 Å ordering band. It is evident that such spectra must be interpreted with care.

An example of a mixed-layer mineral series found in a diagenetic sequence is given in Fig. 3.45 for the illite/smectite series. In this example the proportion of illite and smectite changes with depth changing the position of the individual bands.

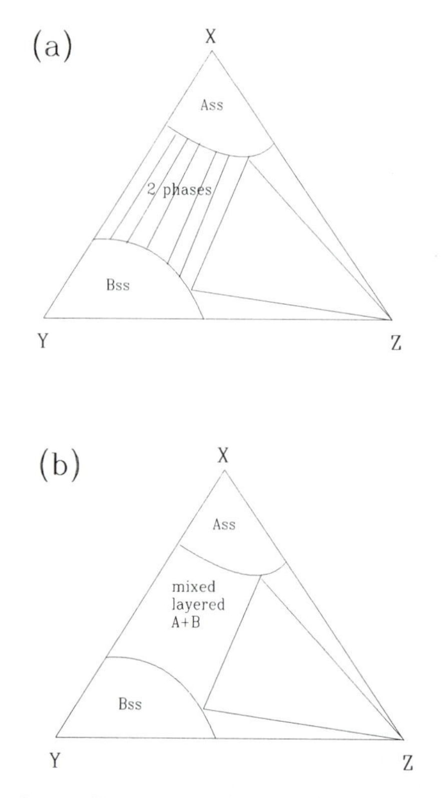

Fig. 3.46 Alternate phase relations according to phase rule criteria depending on how the ordering or interlayering of an interlayered clay structure is arranged. If strong segregation occurs, two phases can be defined and (b) will pertain, if interlayering is random or ordered on a short range scale, (a) will pertain. A and B represent the phases, ss = solid solution phase.

The interesting thing about the mixed-layer minerals is that such combinations provide us with a sort of bridge between the limits of substitution between two different structures. Figure 3.46 illustrates this difference. In the first example, phase A and phase B have a specific limit of solid solution of the components X, Y and Z. Compositions between these phases will give a two-phase assemblage in fixed proportions according to the principle of tie lines. However, in the case of mixed-layer minerals the entire area between the homogeneous phases A and B is occupied by a single, mixed layered phase which contains components of A and B in different successive layers. It is not easy to decide what the status of the mixed-layer mineral really is, as far as the phase rule is concerned. Is it one-phase or two-phase in a special association?

Table 3.14 Mixed-layer clays

Random interlayering	*Ordered interlayering*	*Name*
illite/smectite	paragonite/smectite	Rectorite
glauconite/smectite	illite/smectite	
mica/vermiculite	glauconite/smectite	
mica/dioct chlorite	dioct chlorite/smectite	Tosudite
smectite/chlorite	trioct chlorite/saponite	Corrensite
kaolinite/smectite	trioct mica/vermiculite	Hydrobiotite
mica/chlorite	talc/saponite	Aliettilite
	kaolinite/smectite	

The major mixed-layered minerals found thus far are listed in Table 3.14. Several of the ordered; 1:1 minerals have been given a name indicating their special status.

Mixed-layer minerals usually indicate a change in physical or chemical variables which brings about a change in the stability of one of the clay minerals involved in the mixed layering relative to the other. It is a transition of some sort between two states.

The more common mixed-layer mineral series are illite/smectite and chlorite/smectite which are found in burial diagenesis and hydrothermal alteration events. Muscovite/vermiculite, muscovite/smectite and biotite/saponite or biotite/vermiculite minerals are found in weathering products. Glauconite/smectite minerals are found in sedimentary diagenesis environments.

3.4.8 Oxides and hydroxides

Although most geologists do not usually consider oxides and hydroxides as clay minerals, they do nevertheless have the requisite characteristics of grain size (less than 2 μm) and they do originate at low temperatures. The most common oxides and hydroxides are those of iron and aluminium.

Aluminium hydroxide is the only form of aluminium commonly found in clay-bearing materials. It is called **gibbsite** or **boehmite** (see Table 3.15). These are found in bauxite deposits, those occurrences due to intense weathering or accumulation of the low-solubility aluminium ions.

Gibbsite, $Al(OH)_3$, is found in soils which have been subjected to very intense weathering or in the very early stages of rock decomposition due to weathering. This latter example is probably rare and in any event little gibbsite is produced, being lost in the development of the weathering profile in upper levels. Gibbsite in intensely weathered soils forms in intermediate levels of the profile due to intense leaching of silica and subsequent precipitation of hydrous alumina at lower levels.

Table 3.15 XRD of aluminous hydrates

Gibbsite *d (Å)*	*hkl*	*I*	*Boehmite* *d (Å)*	*hkl*	*I*
4.85	002	320	6.11	020	100
4.37	110	50	3.16	110	65
4.32	200	25	2.34	110	55
3.31	$11\bar{2}$	16	1.98	200	6
2.45	021	25	1.86	$20\bar{2}$	30
2.42	004	20			
2.39	311	25			
2.29	$31\bar{2}$	6			
2.24	022	10			

Table 3.16 XRD Identification of iron oxides and hydroxides

Goethite *d (Å)*	*Lepidocrocite* *d (Å)*	*Hematite* *d (Å)*
4.18	6.25	3.70
2.69	3.29	2.69
2.45	2.47	2.20
	1.97	1.83
		1.69

Boehmite is found in some bauxites, probably owing its origin to diagenesis or low grade metamorphism. It is less hydrated, being an oxyhydrate of the form AlOOH.

Iron oxides and hydroxides are more numerous and more evident in natural clay assemblages. The main colouring agent in clay deposits is due to iron oxides. Colours ranging from yellow through red to brown are due to the presence of these minerals. The minerals goethite, hematite, lepidocrocite and maghemite are the most common forms of iron oxides.

The stability relations of yellow to yellow-brown goethite, α-FeOOH, with the red hematite, Fe_2O_3, are not clear. However, it appears that the effect of pH is very important in the formation of these forms in weathering. Low temperatures, pH near 4 and humid climates favour the formation of goethite. High temperatures, arid climates and pH near 7–8 favour the formation of hematite.

The orange lepidocrocite, γ-FeOOH, appears to form under more reducing conditions than goethite under the same pH and humid conditions.

Table 3.17 Zeolites as a function of increasing alkali content

Mineral	*Si/alkali*
Natrolite	1.5
Analcite	2
Philipsite	2.2
Erionite	2–4
Heulandite	2.2
Clinoptilite	5
Mordenite	5

It is evident that the soils formed under arid or hot conditions will be red in color and the hues will become more yellow and orange as the climate becomes more humid and cooler. The effect of colour on soils due to the presence of these iron oxides is so important that it is a major diagnostic in determining the soil type and giving it a descriptive name.

XRD of these phases in clay aggregates is difficult because the oxides are usually found in low abundance (less than 2%), and they are poorly crystallized or in very small grain sizes. This results in seeing only the most intense XRD bands in powder diffraction spectra. Therefore, only the most intense, characteristic reflections are given in Table 3.16.

3.4.9 Alkali zeolites

This group of minerals, hydrous tectosilicates similar in chemistry to feldspars, are present in clay assemblages in the coarser fractions, near 2 μm. Their presence indicates high silica activity in the chemical environment. The zeolites found at low temperatures, in the common clay facies, are alkali forms, being dominated by K and Na ions. The clays found in zeolite-bearing deposits are dominated by aluminous smectites. Illite or glauconite can be found also but kaolinite is rare.

The chemistry of zeolites is governed by the relation of increasing alkali and decreasing silica content. One can classify the alkali zeolites according to their silica content (Table 3.17).

The thermal stability in diagenesis seems to be a function of the silica content of the zeolites. Those of highest silica content are stable at lowest temperatures whereas analcite and natrolite are found in medium intensity diagenetic assemblages, near 100°C.

The zeolites have a characteristic in common with smectites in that they can exchange cations and anions by absorption. In zeolites, there are channels which can accommodate different ions. Depending upon the zeolite type, the size of the channels varies, giving rise to a high degree of selectivity in the ions absorbed. These give the so-called molecular sieve structures which are used extensively in industry. Initially, natural zeolites were used but more recently synthetic forms have been made on an industrial scale giving rise to 'tailor-made' channels in the structure which are highly selective in their absorption properties. These synthetic minerals are used to select ions from solution and to promote catalysis under higher temperatures than those of clays.

XRD characteristics are difficult to show with ease in that the compositions of the alkali zeolites are highly variable giving rise to very different cell dimensions and to differences in the relative intensities of the bands. There is also a significant amount of overlap in the compositions of different forms, making their identification very difficult. Further, several zeolite species are commonly found together, giving great problems in identification of the minerals. The major diffraction bands are found in the range 2–4 Å where many peaks are found. The alkali zeolites also give diffraction maxima just below 10 Å in the range 6–9 Å. Therefore they are relatively easy to distinguish from clays (phyllosilicates) which have first order intensity maxima in the range 10–15 Å and at 7 Å.

3.5 CLAY MINERAL IDENTIFICATION

In Chapter 2 the methods or tools of investigation in clay mineralogy were briefly outlined. Among these are the basic elements most commonly used to identify clay minerals. Some of the tools are routine, day-to-day methods and others are more useful in difficult problems or those which can give new information concerning the minerals as species. Identification is a day-to-day problem, one based upon well-known and thus slightly false generalizations. All simplification is based upon the assumption that the useful gives way to the slightly inaccurate. In accepting this truism, we can begin to construct a general scheme for identification of clay minerals.

3.5.1 X-ray diffraction

Our first and by far most powerful method is X-ray diffraction (XRD). This method is especially useful in mixed-phase assemblages, the most common situation in routine clay mineral determinations. The range of d planar spacing values can be usefully divided into three areas: those to determine the (001) basal spacings which are in the range 7–30 Å. The 2θ range will depend upon the type of cathode target in the X-ray tube (Cu, Co, etc.). The second region is designed to determine small differences in the basal spacings which are not readily seen in the (001) reflections, by using the (003) reflections. These reflections are usually reasonably intense and the distance between peaks is about 0.05 Å. The third range is near 1.5 Å where the (060) spacings are found that enable the distinction of di- and tri-octahedral minerals and at times the different ionic occupancies in the octahedral sites.

Swelling, expandability (smectites)

The first test is for swelling or expandability of the clays. This is done by running a diffractogram of an orientated specimen (see Chapter 2) in the air-dried state and then another diffractogram in the glycollated state. This is done in the 6–30 Å region.

Comparison of the two diagrams, glycollated and air dried, shows whether or not the diffraction maxima have changed position, demonstrating swelling of the clay. Since we know that the glycol layer increases the interlayer site by 7 Å and that the normal hydrated state of swelling clays is either with a 2.5 or 5 Å beyond the normal 2:1 structural interlayer spacing, it is possible to tell if a clay sample has a fully swelling component in it or one that swells only partially. For example, if the clay changes from 15 Å (two-water layer hydration) to 17 Å there is full swelling but if the change is from 15 Å to only 15.8 Å, there is only partial swelling. This latter behaviour may be due to either an incomplete smectitic interlayer structure or a mixed-layered phase (that is, one containing both swelling and non-swelling layers in the clay particles). In most smectites, the (001) reflection is by far the most intense in the (001) series. Thus it often appears that this is the only band to be affected in the spectrum by the glycol treatment.

Heating the sample to 200°C should expel all of the polar interlayer ions, reducing the 2:1 clays to their basic 10 Å repeat distance.

The comparison of the three diffractograms gives the first classification of swelling (smectite) and non-swelling (non-smectite) clays.

Smectites

The next step is to use the (060) reflection near 1.5 Å to determine the di-tri-octahedral character of the swelling clay. Bands of more than 1.54 Å indicate the presence of a trioctahedral mineral and those near 1.51 Å or less a dioctahedral mineral. This determination allows the major classification of smectites into di- or trioctahedral types: montmorillonite, beidellite and nontronite are examples of dioctahedral clays, while saponite and stevensite are examples of trioctahedral clays.

Heating the prepared sample to 200–300°C will eliminate the glycol and water from the interlayer site of the smectites and a diffractogram taken in the anhydrous state should give a spectrum with a peak at near 10 Å (or slightly below). This indicates a fully expandable mineral. If the peak displacement is less than this, the mineral is either a vermiculite or a mixed-layer mineral.

Non-swelling clays
In those clays which do not change band position with glycollation, the constant peak position gives the clay family. 14 Å chlorites have a constant spacing of 14–14.8 Å. Mica-like minerals (illite, glauconite, celadonite) have a constant spacing near 10 Å. The (060) peak shows illite as having a band below 1.50 Å and glauconite and celadonites above this value. These are the only 10 Å clays stable in the clay environments. However, one must be aware of the possibility of the metastable persistence of muscovite (with (060) at 1.50 Å and biotites with (060) between 1.52 and 1.54 Å). Normally, the small grain size fraction (less than 2 μm) eliminates the majority of the detrital micas.

Kaolinites or 7 Å chlorites (berthiérine, serpentine) have a constant spacing band near 7 Å. The distinction between the two can be made by looking at the (001) band near 3.5 Å where the chlorites have spacing less than 3.55 Å and kaolinites greater than 3.55 Å. The *b* dimension, measured by the (060) spacing shows the distinction between the two minerals on the basis of their octahedral occupancy, greater than 1.51 Å are trioctahedral chlorites and less than 1.51 Å indicate dioctahedral kaolinites.

Sepiolite-palygorskites have constant spacing bands at 12, 7.5, 6.7 Å and 10.5, 6.3 Å respectively.

The XRD method of diagnosis is summarized in Fig. 3.47.

3.5.2 Thermogravimetric analysis

Thermogravimetric analysis is useful to determine the physical properties of the clay in relation to its water content and conditions of release. This is an old method but potentially one of the more useful.

The mineral families can be grouped as before into the smectites, sepiolite-palygorskites and non-expanding phases. The retention of surface water (that which is held on the external crystal surfaces) is similar for all of the clay minerals. It varies only as a function of the surface area of the crystal; smaller crystallites tend to hold more water. The initial loss of this water occurs below 110°C. The bound water of the sepiolite-palygorskites and the smectites leaves the structures between 110°C and 200°C. The crystalline water leaves the structures at different temperatures depending upon the strength of the hydrogen bonds in the crystals. A synoptic table of the ranges of water loss of the various mineral groups is given in Table 3.18.

In general, there are three types of water associated with clays. The surfical layers leave the structures at temperatures below 110°C. The

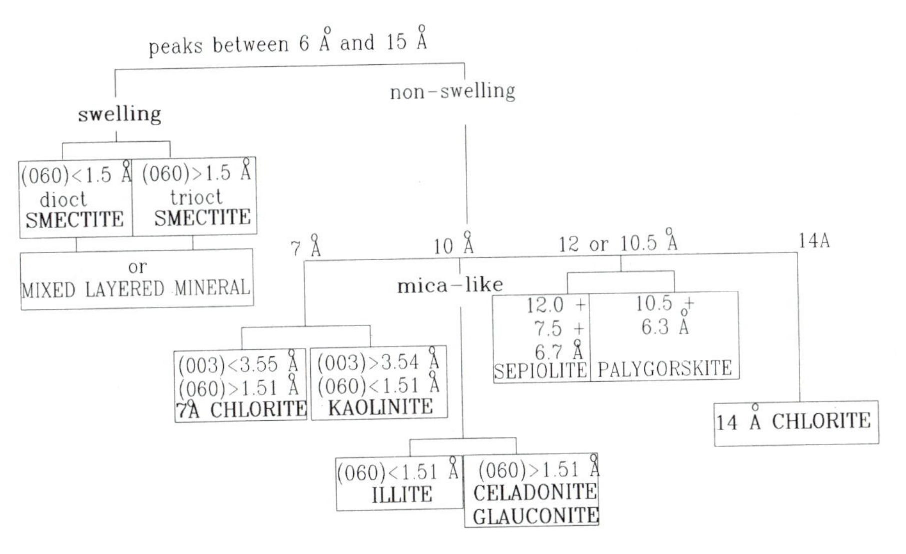

Fig. 3.47 XRD diagnostic for the identification of clay minerals.

Table 3.18 TGA analyses

	Weight loss (%) 10–20°C	*200–300°C*	*500–600°C*	*>600°C*
Smectites				
Montmorillonite, Beidellite, Nontronite, Saponite, Stevensite	12	3	4	–
Vermiculite	6	2	4	–
Sepiolite, Palygorskite	10	2–4	2–3	–
Mica-like				
Illite	1	2	5	–
Glauconite	2	3	12	–
Kaolinite				
Kaolinite	1	–	14	–
Halloysite	7	–	13	–
2:1 + 1 chlorite	–	–	–	8–12
Berthiérine, Chamosite	1–2	–	–	8–12
Talc	–	–	–	6
Pyrophyllite	–	–	5	–

interlayer or zeolite (sepiolite-palygorskite) water leaves the structures at temperatures below 300°C. The crystalline, OH-type water, leaves the structures at temperatures of 400–800°C. The dioctahedral, iron-rich minerals lose this crystalline water at very low temperatures, the aluminous smectites, kaolinites and aluminous mica-like minerals at higher temperatures and the trioctahedral smectites and vermiculite at still higher temperatures. The high-temperature minerals (chlorites, muscovite and talc), which are not clay minerals by their grain size and geological origins, lose this crystalline water at very high temperatures.

The interlayer, exchangeable cations have different hydration energies, that is, bonding strength between the water molecules and the cation. The sequence of temperatures of water loss for the different cations is as follows:

K and Na	<200°C
Ba and Sr	220°C
Ca and Mg	230–250°C

3.5.3 Infrared spectra

Infrared spectral analysis is useful in determining the structural state of a clay and its specific atomic arrangement when single-phase samples are available. To establish that a sample is single-phase, analysis by another method is necessary before IR spectra are established. For this reason, IR is not a routine analysis method.

4

Origin of clays

The origin of clay minerals at or near the earth's surface is, in the majority of cases, a process **instability** in other silicate minerals. It is rare to find clays resulting from aqueous precipitation processes. Most clays are the result of an incongruent dissolution process where the clay is the least soluble portion of the mineral reaction. In processes involving clays at temperatures and pressures above those of the earth's surface, clays are most often the result of **recrystallization** processes. These two major processes, which in fact divide geological origins into two general classes, will be treated separately. A third process is **precipitation** from solution, though this is very rare.

Several specific geological environments are responsible for the crystallization of clay minerals. In all instances the process of clay mineral genesis is accomplished in an aqueous environment. This implies that the aqueous solution has a key role in the formation of clays. This is in fact true. The genesis of clays involves the dissolution of a given mineral or group of minerals, which produces a solution of a different aggregate solute composition from that of each of the reacting solids. From this solution one can precipitate a clay mineral or foster the growth of a clay that has a series of physical and chemical properties which reflect the low-temperature, aqueous environment in which the clay formed. Although the genesis of clays is one of low temperature (between 4°C and about 200°C) the specific geological conditions which led to clay formation are not necessarily contiguous, that is, there are gaps in the series of conditions forming clays.

Basically there are four geological processes which give rise to clay mineral formation: weathering (either subaerial or subaqueous); precipitation from concentrated solutions (saline lakes and closed marine basins); burial diagenesis (effects of chemical and thermal change); and hydrothermal alteration (water–rock interaction at higher temperatures due to thermal effects of magmatic intrusions). All four are water–rock interaction processes. The variables for all of these geological environments are the same, water–rock ratio and rock composition as well as the temperature at which the reactions take place. Essentially this range is from conditions at the earth–air or earth–water interface which gives minimum temperatures of 4°C for ocean bottom formation and around 15°C average for land occurrence to the upper limit at which we find clay minerals, which is near 200°C in young rocks. Burial depths are usually not greater than 6–7 km.

In the four different geological environments mentioned above there are two basic types of physiochemical situation which give rise to the formation of clay minerals: water–rock interactions, that is, those of weathering (low temperature) or hydrothermal alteration (higher temperature); and precipitation of elements from aqueous solution in closed sedimentary basins. These genetic modes will be treated in their geological context.

4.1 MINERAL INSTABILITIES

4.1.1 Types of reaction

Clays are most often the direct product of the transformation of another mineral. That is to say, old minerals dissolve incongruently leaving a solid residue as part of the mineral passes into aqueous solution. This is another way of saying that the interaction between rock and water produces a partial dissolution of the rock, leaving behind a certain quantity of clay. The clay generated during weathering is, among other things, a function of the minerals present in the rock. There are two types of interaction when an old, generally higher-temperature mineral is destabilized: incongruent dissolution of each mineral occurs with the production of a clay specific to the devitalized mineral; or destabilization of two or more minerals can occur with interaction of the low-solubility elements to form a new clay phase which incorporates the insoluble elements of the two initial minerals. These relations can be shown as follows:

$$\text{phase A} \rightarrow \text{clay A} + \text{ions, phase B} \rightarrow \text{clay B} + \text{ions} \quad (1)$$

$$\text{phase A} + \text{phase B} \rightarrow \text{clay C} + \text{ions} \quad (2)$$

Of course one can consider the possibility that clays themselves will become unstable and recombine. This relation can be written as follows:

$$\text{clay A} + \text{clay B} \rightarrow \text{clay C} + \text{ions} \quad (3)$$

Eventually a clay can become unstable and the reaction will be as follows:

$$\text{clay A} \rightarrow \text{oxide} + \text{ions} \quad (4)$$

4.1.2 Hydrolysis

The process of interaction between minerals and aqueous solution is usually one of exchange of hydrogen ions for soluble cations in the old mineral. For example:

$$\text{K-feldspar} + H^+ \rightarrow \text{hydrous clay} + K^+ + \text{Si ions}$$

$$\underset{\text{feldspar}}{3KAlSi_3O_8} + 6H^+ \rightarrow \underset{\text{kaolinite}}{Al_3Si_3O_{10}(OH)_6} + 3K^+ + 6\text{Si ions}$$

These hydrolysis reactions can be used to illustrate the relative stabilities of mineral species as chemical variables by using the aqueous chemical activity (ai = γxi where xi is atomic concentration) of ionic components of the reaction as variables and plotting their slopes for reactions between phases as chemical variables.

In the example given above, three species are found in aqueous solution; K^+, H^+ and silica. In order to represent the variables conveniently, the three values are combined into two, K^+/H^+ and Si. The K^+/H^+ ratio for the reaction is 3 and the Si coefficient is 6. For every three feldspars six silicas are released. The slope of the reaction relation on an activity plot of K/H against Si will be 2. In the reaction as written, three hydrogen ions are consumed to produce kaolinite.

In writing the reactions one considers solids and dissolved species. Some elements are always found in the solids; in the above reaction alumina is always in the solid phase in both reactants and products. This being the case, the solubility of alumina can be neglected, or at least it can be considered to be constant. It is, therefore, important to see which of the low-solubility elements in a system is generally a major part of clay minerals stable as solids. Three key elements in this sense are Fe, Al and Si. Their solubility, absolute and above all relative, is affected by solution pH.

Figure 4.1 indicates schematically the relations of solubility of two of these elements as a function of pH. It is evident that under acid conditions, alumina will remain in the solid state whereas silica will enter the aqueous solution more than alumina. Under neutral conditions both are rather insoluble and under highly alkaline conditions both will be dissolved in a similar manner. In most activity diagrams involving clay minerals which form from acidic or pelitic rocks (those of high alkali and alumina content), the basic assumption is that the aqueous solution is neutral to acid and that, therefore, alumina is highly insoluble or at least much more so than silica. Then variations in silica activity in solution will effect mineral changes. The alumina content of solids will remain stable. Under strongly acidic conditions alumina is more soluble than silica and the assumptions on which the activity diagrams are made must be changed.

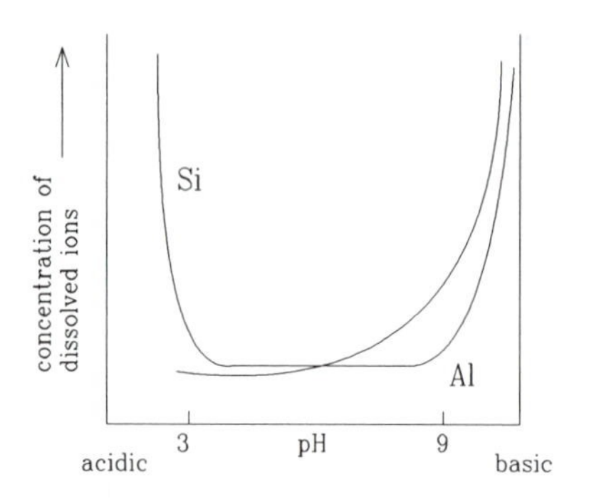

Fig. 4.1 Diagrammatic relations of silica and alumina solubility as a function of aqueous solution pH. At low pH (acid) alumina is soluble while silica is not. Acid soils should then concentrate silica relative to alumina. At high pH (base) both silica and alumina are soluble, which will leave insoluble elements present in the soil such as iron oxide.

For example, soils formed under heavy organic cover tend to show a strong concentration of silica in the upper horizons, demonstrating the effect of high acidity on the removal of alumina relative to silica. This pH effect is enhanced by the complexing action of organic acids which further remove aluminium.

For basic rocks (those which contain less alumina and much Ca and Mg) the elements Ca, Mg, Al and Si are all lost rapidly due to higher pH conditions produced by abundant Mg in solution.

4.1.3 Activity diagrams

Activity diagrams can be used to illustrate the stability relations of the major classes of silicate minerals and their clay destabilization products. **Activity** is the effective chemical impact that a dissolved species has on the solids with which it is in equilibrium. In using such diagrams, and the logic of the chemistry which is the guide line for the diagram, it is assumed that the solids have little impact on the activity of the ions in solution. The solutions are the driving force in the equilibrating procces. Therefore, if one knows the composition of the solution and hence the activity of the ions in solution, one can determine the solids which will be present or in contact (at equilibrium) with the solution. Such diagrams are very useful in water-dominated systems, such as those of weathering or hydrothermal alteration. Figure 4.2 shows five activity diagram relations between clays and non-clay minerals. These relations demonstate the importance of dissolution phenomena and the relations between the magmatic and metamorphic minerals and clay minerals in aqueous, chemical space.

In general, as the water/rock ratio increases, the area of mineral equilibrium will shift to the origin of the activity diagram, that is, lower values of the chemical variables due to dilution. Thus the alteration intensity effect on non-phyllosilicates (production of clays) will be increased as one goes toward the origin of the diagram as aqueous solutions become more dilute.

In the examples shown in Fig. 4.2, there are two types of reaction relations: one where mono- or divalent cations and silica are lost to the solution; and the other where only mono- and divalent cations are lost. In the first system we can see that the relative loss of one ion with respect to another will determine the chemical path and hence minerals produced in the clay forming process. Take Fig. 4.2(a), for example, where potassium feldspar is hydrolysed. If one has an initially high K^+/H^+ activity ratio, dilution of the solution in equilibrium with the feldspar will form mica. At initially lower K^+/H^+ activity ratios and high silica activity, smectite will form upon dilution of the solutions. At still lower ratios kaolinite forms directly from potassium feldspar.

In Fig. 4.2(b) the activity relations for albite are shown. Here it is apparent not only that one can produce clays (kaolinite), but also that a new mineral type is present, the zeolite analcite. The potassic and sodic systems then show a significant difference in the sequence of alteration phases produced.

In a calcic system, which has magnesium as the relatively insoluble element (Fig. 4.2(c)), the high-temperature and unstable mineral is diopside which is transformed into amphibole (tremolite), talc or serpentine (antigorite). These last two minerals are not precisely the clay phases one would find in most geological situations but they can be assimilated to saponite, vermiculite and the clay chlorites which are the more common phases likely to be present.

In order to include more common high-temperature phases in an activity diagram, one can assume that alumina is an insoluble, inert element and then look at the effect of increasing water/rock ratio by choosing the ratios of magnesium and potassium ion activities to hydrogen ion activity. In such a diagram (Fig. 4.2(d)) one can find the common granitic rock minerals potassium feldspar, phlogopite and muscovite. The new clay minerals will be kaolinite and chlorite.

A representation of magnesium and aluminium activity compared to hydrogen ion activity with silica insoluble (Fig. 4.2(e)) can be used to indicate the change of clay minerals into other clays or oxides (quartz). Here the initial rock would contain kaolinite serpentine and aluminous chlorite which will be destabilized to talc (probably saponites) and eventually quartz alone.

These diagrams are only schematic in their relations concerning specific

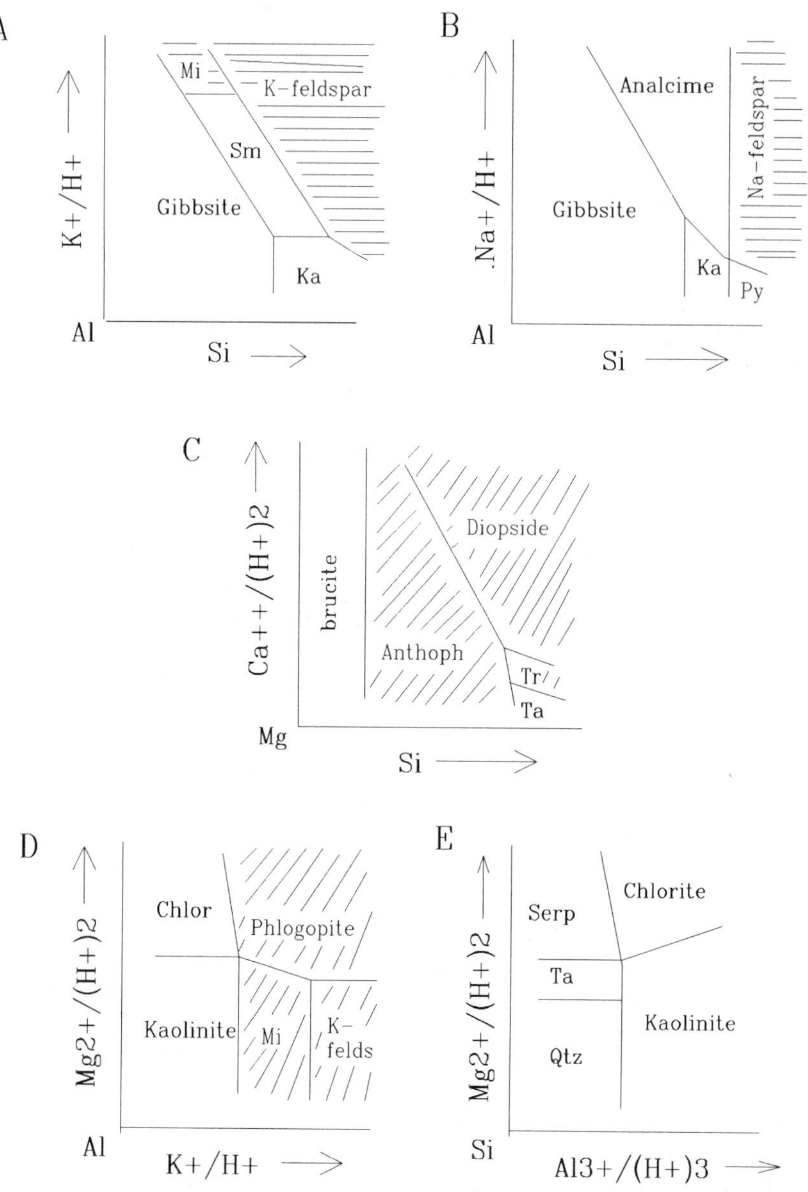

Fig. 4.2 Activity diagrams showing the stability relations of different high-temperature minerals and clays as a function of pH (H^+) and dissolved ions (M^+). One element is considered insoluble. It is found at the origin of the diagrams. (a) Stability of potassium feldspar (K-feldspar) as a function of the relative activities of hydrogen ion, potassium ion and aqueous silica. Alumina is insoluble.

Fig. 4.2 *(cont'd)* High-temperature feldspar is destabilized by a decrease in either K^+/H^+ or Si activity while muscovite (Mi) is destabilized by a decrease in the ratio of K^+/H^+. Smectite (Sm) and Kaolinite (Ka) replace the muscovite and feldspar phases (lined area). Gibbsite (Gi) is produced by extreme chemical conditions, those of very low concentrations of silica in solution. (b) Stability of sodium feldspar (albite) as a function of the activity of Na^+/H^+ and silica. Note the reduced range of conditions which stabilize kaolinite. Much of the clay field is taken up by the zeolite analcite (Anal). Py = pyrophyllite. Al is insoluble and its activity is considered to be constant. (c) Diopside-amphibole minerals (shaded areas) are changed into clays as a function of the activity of $Ca^{2+}/(H^+)^2$ and silica. Magnesium activity is constant and Mg is considered insoluble. The talc field (lined area) is restricted to high silica and low calcium activities. Bru = brucite, the hydrous magnesium oxide; Anth = anthophyllite; and Tr = tremolite; Diop = diopside. (d) Relative stabilities of di- and trioctahedral micas (Mu = muscovite, Phlog = phlogopite) and potassium feldspar (K-felds). This is a double ratio diagram where hydrogen ion activity is included on both axes. Kaolinite (Ka) and chlorite (Chl) occur at low potassium and magnesium ion activities. Aluminium activity is considered to be constant. Phlogopite, muscovite and potassium feldspar (shaded area) occur at higher potassium ion activities. (e) A clay–clay system illustrates the relations of aluminous, 14 Å chlorite (Chl), 7 Å serpentine (Serp), talc (Ta) and kaolinite (Ka) at constant silica activity. Low Al and Mg activity give rise to quartz (Qtz) as silica is considered to be insoluble.
Diagrams based on data from Bowers *et al.* (1984).

clay minerals. One reason is that not enough reliable thermodynamic information on all of the clay phases is available to construct a decent phase diagram based upon activity relations. A major problem is that most clays, kaolinite excepted, have a wide range of compositions which make both determination of their thermodynamic values and their representation difficult on simplified coordinates. A determination of their thermodynamic parameters must often be dealt with in an indirect manner such as by using model calculations. At the moment one cannot deal with these problems simply but the diagrams presented give a general indication of the mineral relations in changing chemical, aqueous solution conditions.

Figure 4.2 shows that the hydrolysis of metamorphic or igneous minerals has the effect of decreasing the alkali or alkaline earth element content of the solids and increasing their alumina content, assuming neutral to acid pH conditions.

Another factor, rarely discussed when dealing with silicates, is the effect of oxidation potential. For example, in weathering, the ferrous iron held

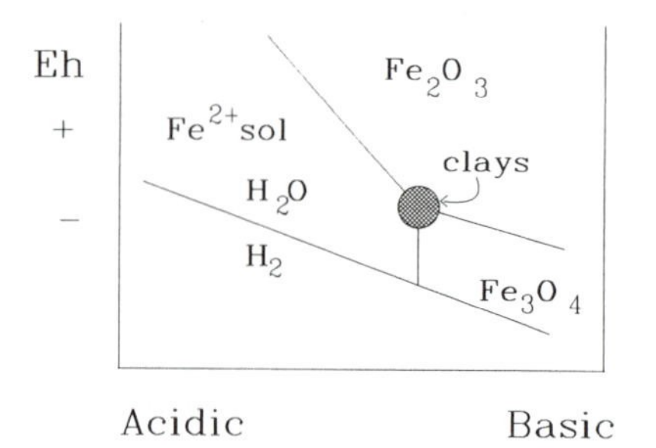

Fig. 4.3 Relations of iron oxide phases under various conditions of Eh and pH. The H_2O-H_2 line shows the lower limit of water stability. Acid and high Eh conditions favour aqueous divalent iron, basic and reducing conditions favour the mixed-valence oxide magnetite (Fe_3O_4) and basic oxidizing conditions favour hematite (Fe_2O_3). In silicate systems, magnetite could be replaced by ferrous-ferric clays, and hematite could be replaced by ferric clay minerals. The zone critical to the type of clay which will form is at slightly alkaline and reducing conditions (shaded) where the oxidation and physical state of iron will change. It can be seen that acid conditions favour dissolution of iron-bearing phases, higher pH and oxidation potential will concentrate iron oxide, while basic solutions will induce the presence of oxidized iron in the solid phases (silicates). Eventually ferric silicates form and these are destabilized at higher effective oxygen activity (more oxidizing conditions) to form ferric oxides.

in high-temperature silicates is typically transferred to clays first as ferrous ions and then these ferrous clays in turn are transformed into ferric clay types. Ferric clays may in turn be destabilized to oxides. The oxidation state of iron determines the type of clay mineral which will be present in the soil.

The relations of iron oxides and dissolved iron species as a function of pH and oxidation potential (Eh) are discussed at length by Garrels (1960). A representation of the relations which directly concern the stability of clays is given in Fig. 4.3. The low solubility of Fe^{3+} ions in aqueous solution leads to the formation of a ferric oxide such as hematite. Thus the limit between the stability of the aqueous Fe^{2+} ion and ferric iron oxide will be critical in determining the clay assemblage which will form during the destabilization of a high-temperature iron-bearing mineral. If the Eh–pH conditions favour ferrous iron, trioctahedral minerals (saponites and talc) will form. If the solutions favour ferric iron, dioctahedral ferric minerals (nontronite) will form or oxides will be present instead of silicate clays. In cases where the mixed valence oxide magnetite is stable, Fe^{2+}-Fe^{3+} clays (berthiérine) will form.

4.1.4 Alteration of phyllosilicates

The transformation of high-temperature phyllosilicates into clay minerals often proceeds in a different manner than that of non-phyllosilicate phases. This is probably because of a strong similarity in the structure of the unstable phase relative to the new one. Change from phyllosilicate to clay is often noted to be progressive, that is, to proceed in steps. In the case of biotites, for example, the change from mica to smectite is effected first by the formation of an ordered mixed-layer phase containing smectite-biotite units alternating in a regular sequence. This is often called **hydrobiotite**. The chlorites from metamorphic rocks can be observed to form a similar mineral type, smectite/chlorite mixed-layer phase. The chlorite to corrensite (1:1 chlorite-saponite structure) is quite commonly noted in soils. Muscovite and illite have also been observed to be transformed into regular interlayer smectite-illites. Many of the soil clay minerals formed by the destabilization of phyllosilicates form a mineral known as **vermiculite**. This phase is ill defined, and probably represents various stages in the change from non-expanding to expanding minerals.

Further weathering (completion of the transformation process) changes the structures entirely to the new, low-alkali, calcium- or magnesium-bearing, expandable clay minerals.

4.1.5 Geological environments

The origin of clays is very closely related to the geological environment in which they form. The conditions of temperature and chemistry are the factors which determine the type of clay which will be present. The geological environments where clays are found can be divided into the following catagories: surface either air-dominated (that is, continental) or subaqueous. In both of these environments one can talk of weathering in that there is an influence of high quantities of aqueous fluids which exchange some components with the rocks which are affected. The differences are the relative amounts of liquids involved and the content of dissolved salts. The next environment is that of **burial** or continued sedimentation. The effects of a more confined chemical system and an increase in temperature affect what is called **diagenesis**. Another environment is one of water–rock interaction at high temperature, **hydrothermal alteration**. In these situations, most often associated with the intrusion of a magma of some type, the clays are found to form over a range of chemical and thermal conditions in a relatively restricted space. Veins show rapidly changing physico-chemical conditions. These characteristics are found in continental settings and in rocks found on the ocean floor.

The discussion which follows gives an insight into the clay types formed as a function of the geological environment or their origin.

Origin of clays by subaerial weathering

Among the geological environments where clays form and are stable, the weathering site is most important for the activities of man. It is at the rock–surface (atmosphere) interface that man's activities come into contact with nature to the greatest extent. Acid rain, chemical waste, farming and the pesticide-fertilizer cycle all pass through the soil–atmosphere interface where clays are found. Thus an understanding of the clay-forming processes and the properties of clays found in soils is crucial to understanding the problems of the environment. In order to understand the systems which affect clays in soils and weathering environments it is necessary to understand the physical-chemical variables operating here.

There are several factors which dominate the interaction of surface waters with rocks in the weathering process which produces clay minerals. The factors can be considered as being chemical or physical in nature.

In weathering, the normal parameters considered by geologists and soil chemists are rock type, climate, topography and age. These four variables lead, normally, to the development of a specific type of vegetation at the surface of the profile which then plays an important role in the further development of the weathering. The four variables can be broken down into more conventional chemical or physical variables.

Basic factors in weathering

Four geological or geomorphological variables can be considered. The first is **rock type**, which is a chemical factor. The second is **climate**, which is composed of rainfall (a chemical factor) and temperature (a physical factor). The third is **topography**, or **flow rate**, which is a chemical factor because it determines the ratio of water to rock through drainage. This drainage can be largely conditioned by the permeability of the soil or altering rock. In general, weathering in a swamp is very different from that on a steep hillside. The fourth variable is the **age** of a profile or weathering sequence, which is a physical parameter (time).

Of the basic factors – chemistry, temperature, time – the chemical factors of rock composition and water/rock ratio are probably the most important in determining the type of clay mineral which will be produced. Temperature and time are kinetic factors which determine the rate at which the chemical process will proceed.

An illustration of one of these parameters is given in Fig. 4.4, a typical plot of clay content of the soil versus rainfall. On this plot the different clay mineral types are given as a function of the two variables. Clay

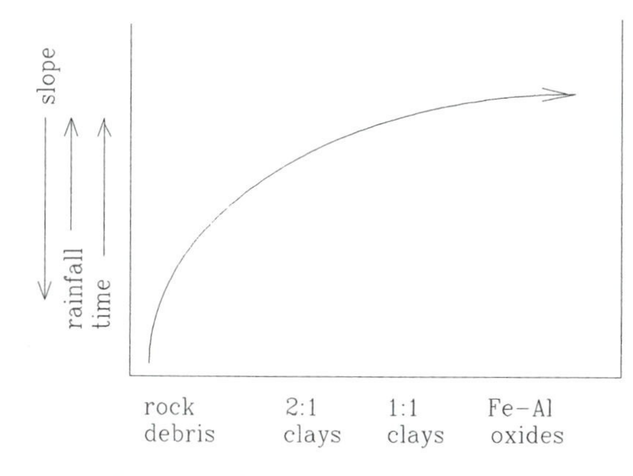

Fig. 4.4 Relations of clay mineral types formed under reaction intensity variables. Time and rainfall affect the reaction variables which change the clay type present in weathering profiles. Both variables produce the same effect. Decrease in slope slows reaction progress.

mineral content is a rough measure of reaction progress in the transformation of minerals to produce clays. The rainfall coordinate (a chemical parameter) indicates hydrolysis intensity as effected by the soluble ion/H^+ ratio. In the figure it can be seen that the greater the rainfall, the greater the clay content. The amount of rainfall (water/rock ratio) is important in that it defines the approach to equilibrium between rock (silicates) and aqueous solution. As a solution is brought into contact with a silicate-bearing rock, the chemical reactions proceed towards an overall equilibrium between the solids and the liquid. The more solution there is, the more the equilibrium will be brought to its chemical state (pH, dissolved species, etc.). In the reverse case, the more rock there is, the more the solution will be in equilibrium with the rock. If there is a great predominance of rock (low water/rock ratio) the solution will not affect the silicates in the rock. Therefore, the rainfall indicates the water/rock ratio and this indicates whether the rock will try to come to equilibrium with the solution or whether the solution will be trying to come to equilibrium with the rock. In the latter case, the amount of alteration (formation of clays) will be small.

If we consider rainfall and the length of time a rock has been exposed to it, one can combine these reaction variables of weathering into hydrolysis intensity (chemical) and reaction advancement (time) coordinates. Figure 4.5 indicates the broad types of clay mineral which can be expected to form as a function of reaction progress and hydrolysis intensity which is really the amount of exchange of hydrogen ions for K, Na, Ca and Mg ions lost during water–rock contact. In the diagrams of Figure 4.2, the

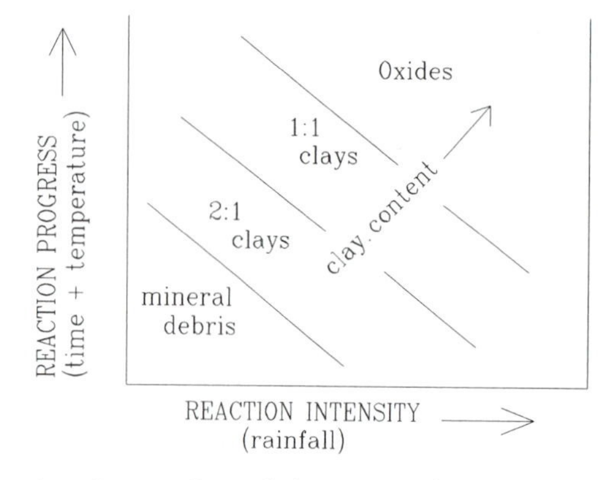

Fig. 4.5 Clay mineral types formed during weathering as a function of reaction progress of the weathering process. More clay is formed at higher rainfall levels (more dilute aqueous solutions) and as more clay is present the mineral type changes from smectite (2:1 structure) to kaolinite (1:1 structure) and eventually only hydrous oxides of aluminium and iron are present at high rainfall (lowest silicate ion activity in the solutions). The sequence demonstrates a loss of silica from the clay assemblages.

increase of the advancement-intensity parameters is seen in a movement toward the origin in the figures.

Figure 4.5 can also be read, again in a very general way, as a representation of a soil profile. In the upper region, where the water/rock ratio is greatest due to solution renewal and dilution by rainfall, the clays tend to be 1:1 types with ferric iron oxides, often hydrated, or perhaps alumina in the form of gibbsite. As one goes down the soil profile, the residence time of water intimately related with the rock increases and the effective water/rock ratio decreases. Dilution of the solutions is less important. One finds more 2:1 clay minerals in this portion of the profile. Finally, in the lower reaches of the profile where the rock predominates over the water content, unreacted minerals are the norm.

Figure 4.5 can be interpreted in terms of climate by looking at the hydrolysis intensity axis as a function of rainfall, increasing intensity corresponding to more rainfall, and along the reaction progress axis as a function of temperature. Older profiles will follow the same axis whereas younger ones will be found towards the origin of the diagram. Obviously, if one considers different combinations of the variables of weathering one will need more axes.

Slope is very important. The rate of drainage or flow rate is determined by topographic and permeability factors (soil structure). The higher the flow rate, the less time the solutions are in contact with the rock and the greater is their ability to remove ions from the solids. They are never in equilibrium with the rock nor saturated with the elements in the solids. Higher flow rates show low interaction coefficients and fewer clays produced.

Basic structure of a soil profile

In order to understand the interaction of rock and aqueous solution, it is important to know the basic structure of a soil profile developed upon a basement rock. We will initially not consider soil profiles developed upon sediments because they frequently involve the transformation of pre-existing clay assemblages and one sees only clay–clay transitions. The basic starting unit considered in this section is the crystalline rock. In such rocks, all of the minerals become unstable and are replaced by clays and ions lost to the percolating aqueous solution. Such a situation more clearly shows the importance of clay mineral formation.

In Fig. 4.6 several comparisons are made: the percentage clay and cation exchange capacity of the horizon are indicated as a function of relative depth. An indication is given as to the nomenclature of the horizon in the alteration profile. This last indication is quite global. The problem of nomenclature of soil profile horizons is very complex and at present there is no general consensus among the many authorities. Therefore, only general categories are given to indicate approximate positions in soil profiles.

Rock In order to follow the sequence of clay genesis as a function of reaction progress, we will consider the initial phases of water–rock interaction and move up the alteration profile progressively. The first zone, then, is that of the initial unaltered rock. In all rocks which are found at the earth's surface, there are many cracks and fissures due to the release of thermal and pressure constraints produced at depth. These cracks and fissures are the initial paths of penetration of the changing fluid. The longer the crack or fissure, the greater the depth to which it penetrates the rock, and the greater the alteration which will occur along it because it will carry more water. The more water it carries, the wider the crack will become by dissolution of the silicate phases. As the ratio of water to rock remains high, the alteration zone on either side of the rock will develop to a greater extent by incongruent and congruent dissolution of the silicates. The initial cracks in a rock exposed to surface alteration are the pathways of the transformation and thus clay-forming processes. The sequence of reactions forming the clays will be seen to proceed outwards from these initial pathways. When smaller cracks become active, and even smaller ones enter into the alteration sequence, the network of alteration will become such that most of the minerals will be altered and the rock will be

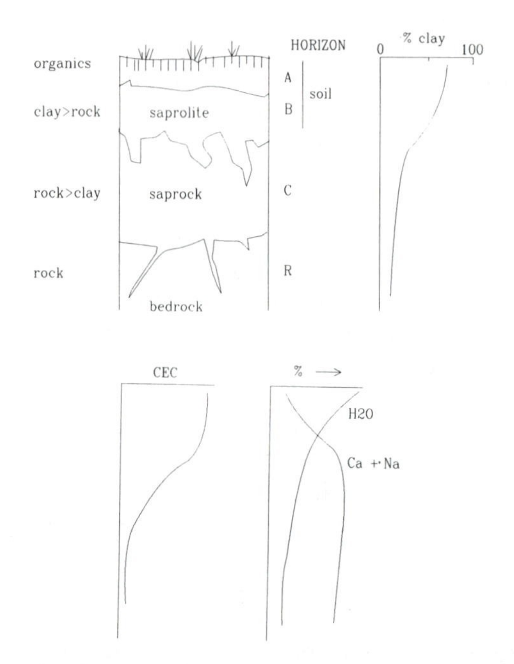

Fig. 4.6 General relations of weathering on crystalline rocks. The double nomenclature of soil-rock type and soil profile horizon (A to R) is compared. Changes in the clay content, cation exchange capacity (CEC) and chemical content of the horizons are shown.

homogeneously altered (transformed into a mixture of clays and oxides), but in the initial stages of alteration one finds zones near fissures (cracks) which show intense alteration and intermediate zones with intermediate stages of alteration while the zones furthest from the fissures will be unaltered and will not contain clays.

The alteration structure in the initial rock is then inhomogeneous when one compares crack areas to those which are relatively fissure-free. This alteration structure is suggested in Fig. 4.7. Large fissures will have significant alteration halos where the initial phases of mineral transformation have begun throughout the minerals. At the edges of these halos reaction occurs mainly at contacts between mineral grains. In many parts of the rock, there is no reaction visible. Thus in the initial stages of weathering a sample is heterogeneously altered. In the upper horizon of the bedrock, one often finds initial formation of clay minerals localized around the initial cracks in the rock. Weathering has already commenced.

Saprock (C soil horizon) The first alteration zone is called **saprock** (C soil or weathering horizon). In this horizon the initial minerals are more or less altered to clays but the petrographic structure of the rock,

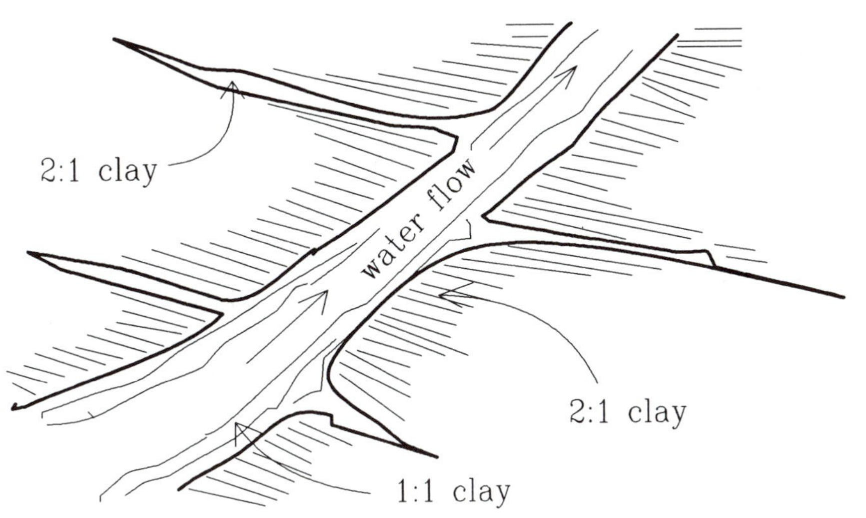

Fig. 4.7 Schematic diagram of alteration intensity around a fracture in a macrocrystalline rock. The smaller, narrower fracture zones produce 2:1 type clays which are more dominated in their composition by the minerals which the crack traverses. The larger fracture zone shows a dissolved central zone, created by high flux of solution, which is now coated with a secondary clay deposit that has been transported to its present site. This is often a 1:1 clay such as kaolinite. The walls of the large fracture are highly altered into a predominantly 2:1 type clay. This illustration shows that two different clay types can be found in the same rock alteration zone depending upon their position relative to the influx of altering solutions.

that is, the location and size of the initial mineral grains, is largely respected, as is the chemical imprint of the initial minerals of the rock. For example, one can see the white areas where feldspars were present and darker zones where ferromagnesian phases (pyroxenes, amphiboles) occurred, although these phases may have largely disappeared. The rock is microscopically very complex. One can usually see several types of clay mineral in a small surface area under a petrographic microscope. The alteration structure is controlled by the initial rock fractures which at this level have become major pathways for the movement of solutions. Along these fissures and fractures the clay-forming reactions are decidedly more advanced than in areas further from them.

The initiation of an important phenomenon occurs in this zone. It is the transportation of clays in the percolating fluids by physical displacement. Along cracks and pores one begins to see clay mineral coatings which are orientated parallel to the walls of the passage. The origin of these clays is the upper alteration horizons and thus they do not represent chemical equilibria in the saprock. However, they can influence the chemistry of the solutions which follow the pathways that they are coating.

The saprock clay and mineral structure is, then, very heterogeneous. In this zone the overall quantity of clays is greatly increased over that of the basement rock through the alteration and transformation of the original minerals. The most important feature of this horizon is the extreme heterogeneity of the mineralogy with many types of clays and large portions of unaltered or little altered rock minerals.

The clays which are produced most often respond to very local chemical environments, either at a contact between mineral grains or within a single grain itself. This produces clay minerals whose composition and mineral type depend entirely upon the initial rock mineral itself and not upon the rock as a whole. In one small area, several millimetres square, it is possible to find clay minerals which would not be compatible if they existed in the same chemical system. For example, clays formed in a feldspar will not be in equilibrium with those in a pyroxene which is contiguous with it.

The reactions in the saprock, C horizon are of type

$$\text{A} + \text{B} \rightarrow \text{clay} + \text{soluble ions}$$

or

$$\text{A} \rightarrow \text{clay} + \text{soluble ions}$$

In upper levels of the C horizon some clays can become unstable and reactions of type

$$\text{clay A} \rightarrow \text{clay C} + \text{ions}$$

or

$$\text{clay A} + \text{clay B} \rightarrow \text{Clay C} + \text{ions}$$

will occur. Again this horizon is typified by an increasing heterogeneity of clay types. The stage of reaction advancement is irregular in the horizon, depending upon the relative amount of fluid which circulates relative to the rock surface with which it is in contact.

This is the zone where clays form from incongruent dissolution of high-temperature phases.

Saprolite (B weathering horizon) The next zone is that of the saprolite (horizon B), where there is little or no rock structure left but where the chemical segregations due to the initial mineral grains are somewhat intact. The upper portion of the saprolite zone is part of the soil horizons. Clays predominate.

The effect of chemical transformation is greatest here; most original minerals are altered to a large degree. Physical transportation of clays (illutriation or lessivage) from the upper soil horizon along fracture and pore paths is also important. The clays of the saprolite zone can have two origins: one chemical, determined by local chemical reactions; and another due to mechanical transport. The clay minerals present in this horizon are not all necessarily formed under the same chemical conditions.

Due to a high degree of mineral transformation in the upper part of the saprolite zone, there is frequently a collapse of the rock structure; old fractures and fissures are lost. There is a general tendency for the clay minerals to be physically intermixed in this zone and as a result one finds new clay minerals forming from old ones due to the new chemical equilibria established by the physical mixing operation. Here clay–clay reactions are very important. The new clay phase most encountered in temperate climate weathering is vermiculite.

The B horizon is characterized by a very high proportion of clay minerals. This is the zone of the initiation of clay–clay reactions in a basically clay matrix.

Surface (organo-mineral A horizon) The uppermost weathering horizon commonly contains much of what is called the soil. This is where, in a natural and undisturbed state, the uppermost superficial layer is dominated by organic matter. In most areas the activity of agricultural man has tended to mix the upper part of this horizon with the one immediately underneath it and at times with the saprolite level further down. In an undisturbed and well-developed soil the uppermost horizon is often characterized by low clay content due to illutriation of clay minerals by infiltrating solutions. The minerals present tend to be those insoluble in aqueous solution such as zircons, iron and titanium oxides. New minerals produced by weathering are often hydrous oxides of iron and aluminium or the dissolution-resistant mineral quartz (SiO_2).

Just below the organic layer of the soil horizon there is an accumulation of oxides in well-developed soil profiles. Under acidic, organic-rich conditions one finds silica, in such forms as quartz sand, whereas in others (neutral to basic) there is an accumulation of alumina and iron in the form of hydrous oxides such hematite and gibbsite. Alumino-silicate clay minerals can be almost absent due either to chemical instability or due to mechanical transport through illutriation. The clays are swept to lower horizons by the passage of the altering aqueous solutions. They deposited on pore walls or in cracks in the lower horizons.

In the A horizon the chemical and physical properties of the system are dominated by the pore structure. This is a secondary structure due to the weathering transformations which have totally effaced any rock structure. Both the soil and the upper saprolite horizon are characterized by clay–clay and clay–oxide transformations.

In the upper soil horizon oxides are produced to the detriment of clay minerals.

In most weathering profiles several major characteristics can be noted. In general, the soil clay mineralogy is more complex (more different species present) as one moves up the profile. This indicates that surface alteration profiles will be signally multiphase. The general trend in bulk composition of the materials in each horizon is that of a loss of first Na_2O, then CaO and MgO. Al_2O_3 and SiO_2 tend to be concentrated. When iron is oxidized, the subsequent destabilization of the silicates augments the loss of divalent ions in the ferrous silicate structures. This transformation of oxidation state of iron leads to an abundance of trivalent ions (Fe^{3+} and Al^{3+}) in the silicates as well as in oxides. One finds kaolinite (Al^{3+}) and nontronite (Fe^{3+}) in the upper levels of many weathering profiles. Extreme chemical action, low pH or very high water/rock ratios due to heavy rainfall, produce oxide concentrations such as Al_2O_3 and Fe_2O_3 with residual quartz (SiO_2) to the detriment of silicates.

Effect of climate

Climate, as it affects the weathering process, is a variation of temperature and rainfall. Various combinations of these two factors give different climates. Thus one can have hot and dry (desert), hot and wet (tropical) or cold and wet (arctic) as well as any intermediate combination of conditions. The variables are those of chemical and physical conditions (temperature is the factor of reaction speed). The following is a generalization of different major climatic types.

The intensity of weathering is a function of the two variables rainfall (chemical) and temperature (physical). These variables affect the relative development of the different parts or horizons of the profiles. Figure 4.8 indicates the relative development of the different horizons (saprock, saprolite, soil or C, B, A (see Section 4.1.2)) in typical soils of different climates.

Desert and arctic conditions

Under climates of low chemical reactivity and little reaction advancement, one finds little weathering activity. In the altered zones there is a greater development of the A and C horizons. The clay-rich B horizon is greatly reduced.

These profiles are typical of arctic (low temperature, a physical variable) and desert profiles (low water/rock ratio, a chemical variable). In these types of weathering mechanical actions are predominant. The thermal effects of heating (desert) or freezing (arctic) produce smaller grains of the original minerals through the action of differential thermal expansion of the solids (stress developed by changes in the relative sizes of individual grains due to anisotropic thermal expansion) or solids and liquids (ice

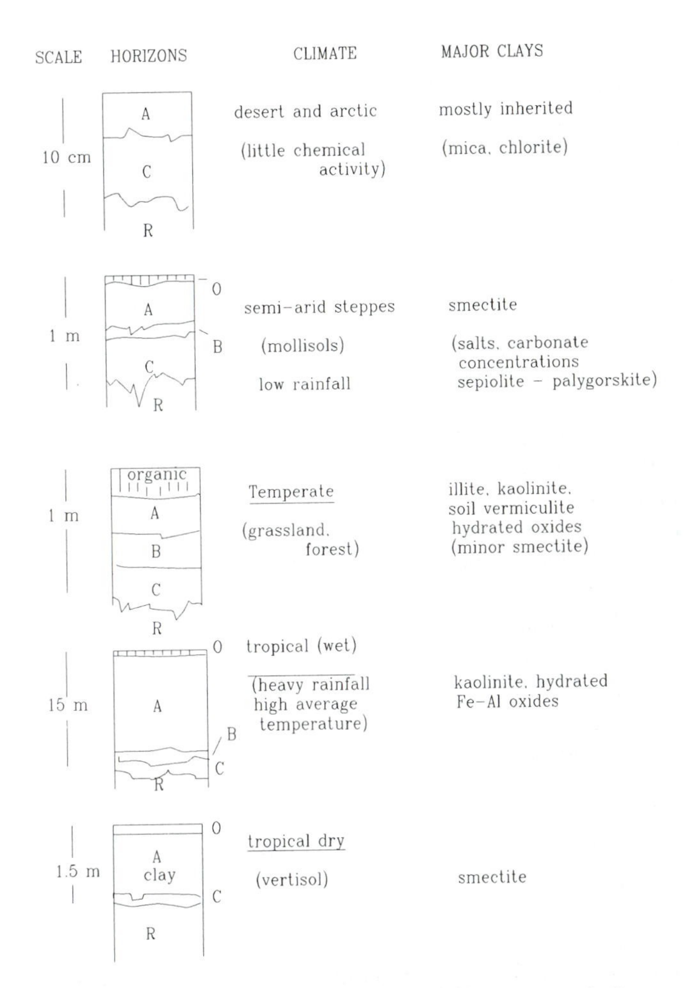

Fig. 4.8 Representation of weathering profiles in different general climates. Major clay types formed under these conditions are shown. The shaded area at the top of the profile indicates the thickness of organic matter in the profile. The A horizon, clay rich, is generally well developed in all profiles whereas the C and B horizon development is greatly affected by the climatic conditions (rainfall and temperature).

wedging). Clays are formed due to the limited chemical activity of the rock–water associations (either by low temperatures or low water/rock ratios). Those clays present are most often inherited from the source rock or sediment. Thus in these geological surface environments, the rocks are disaggregated to smaller grain size aggregates but little chemical activity has occurred and few new phases are formed.

Arid steppes and high plains

In areas of more or less clement temperatures, contrasted seasons but low rainfall, such as on high plains and steppes, one finds weathering which produces a soil type known as **mollisols**. This weathering type typically gives smectites as a dominant clay phase. The B horizon is little developed.

In soil profiles developed under more arid conditions there are accumulations of sulphates, carbonates or salts in the small B and C horizons depending upon the composition of the source rocks or sediments. The carbonates are frequently accompanied by sepiolite-palygorskites. The organic layer at the surface is of variable depth. In these semi-arid weathering climates, the sparse rainfall enters the upper regions of the profile but subsequent dry conditions move this humidity back upwards in the profile by capillary action. As the solutions move towards the surface, they are concentrated by evaporation, whereupon they react with the solids under chemical conditions different from those of initial rainfall. The upper regions are those of chemical precipitation of dissolved salts, sulphates, carbonates and oxides, and the formation of sepiolite and palygorskite.

Temperate climate

Under climates conducive to the development of heavy vegetation, grasslands and forests, one finds profiles with significant development of all weathering horizons. The organic soil zone is actively developed as well as the B horizon. Illite and kaolinite are the dominant minerals in the weathering profiles when aluminium is present in sufficient quantities. In the A horizon, when high organic activity lowers the pH sufficiently, one finds a development of silica (quartz) just below it. This is typical of the podzol soil type.

Tropical dry climate

Conditions of high average temperatures and uncontrasted climate under dry conditions produce the vertisol weathering type. This is dominated by the presence of smectite forming from almost any rock type. The high expandability of the smectite gives a special character to the weathered zones. In dry periods large cracks develop to great depths due to the significant loss in volume of the smectite. Brief rainfall quickly changes the soil structure to one of heaving and perturbation. The A horizon is greatly developed. It is dominated by clays.

Tropical humid conditions

In tropical, rainy climates the A horizon is very greatly developed and the B and C portions are very limited in thickness. The A horizon is dominated by hydrous oxides of iron and aluminium. The high rainfall and warm temperatures promote the development of 1:1 clay minerals

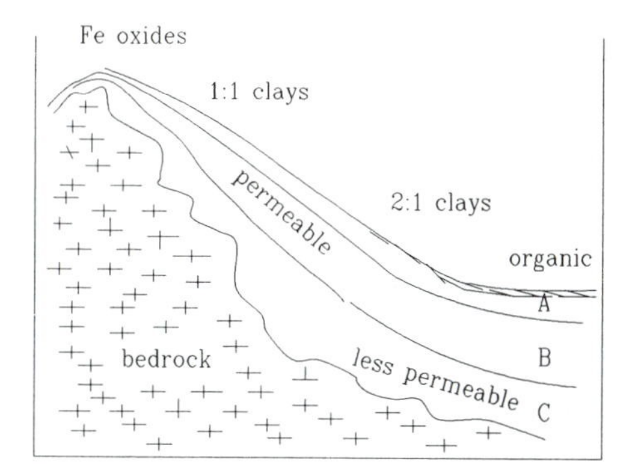

Fig. 4.9 Illustration of the importance of slope in the development of the different alteration zones under temperate climatic conditions. In general, the higher slopes favour the development of oxides and mineral debris; lower slope gradients give rise to the predominance of 2:1 minerals. The chemical effect is determined by the residence time of solutions in contact with the rock. This determines the solution composition which is most important in forming the clay type in the soil profile.

and oxides which form rapidly after the bedrock has been initially altered into 2:1 clay minerals (B and C horizons).

Effect of slope

Rock slope gradient contributes to the type of profile developed. This effect is very similar to that of climate in that the slope controls the water/rock ratio and hence the chemistry of the reactions. The greater the slope, the more rapid the displacement of the aqueous solution over the soil profile. This short residence time does not allow extensive water–rock interaction and the chemistry appears to be one of a high water/rock ratio. Soils on a steep slope tend to have little developed clay zones (A, B horizons). As the slope decreases, the clay-bearing portions of the weathering profile are increased. This is represented in Fig. 4.9 for a weathering sequence in a tropical climate zone. The overall weathered zone is least important near the summit of a slope and the lowest portions of the slope show deepest weathering effects.

Also, the general type of clay mineralogy changes with the slope. In the upper areas, where effective water/rock ratios are great, oxides of iron and aluminium tend to form. As one proceeds downslope, 1:1 minerals (principally kaolinite) are common, and in the lower regions one finds 2:1 minerals such as smectites forming.

The sequence shown in Fig. 4.9 can be considered as a sequence of soil types related by differences in slope under the same climatic conditions.

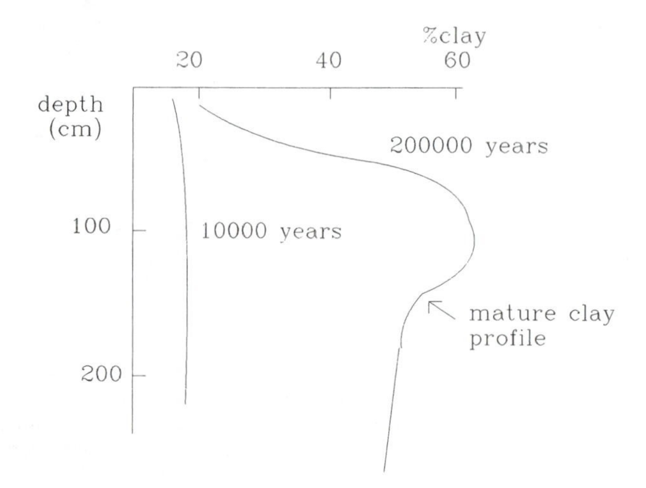

Fig. 4.10 Importance of time in the production of clay minerals. The older profile shows a greater clay content than the younger profile.

Effect of age

The length of time of weathering is important as it determines the amount of clay produced. The transformation of rock into soil is a function of the time of the reaction process, as are all chemical reactions. Young soils show little reaction and the products of this process are sparse. Physical disaggregation processes are more important in producing smaller elements in the profile than are chemical ones. As time increases the clay-forming processes are more developed and the depth to which the clay-forming reactions have occurred is greater.

Many attempts have been made to determine the importance of time in the weathering process. The major problem is in finding profiles developed on similar materials which can be dated. The easiest are river terraces and sedimentary materials which include dating devices in them (fossils, terrace levels, glacial ages). Granites, for example, are poor subject matter for such studies as they give no clue as to outside geological events.

The differences in weathering rate found by various workers in different geological settings, mainly in glacial sediments or river sediments and terraces, can be attributed for the most part to the differences in the physical and chemical properties of the materials which are altered. One example is given in Fig. 4.10, where a soil profile is developed on glacial till (finely ground rock which is slightly weathered to the clay assemblage illite + chlorite in many cases) in a temperate climate in Pennsylvania, USA. The weathering profile developed after 10 000 years shows little change in the clay content. Thus one can suspect that little transformation has occurred. The profile produced after 200 000 years of alteration shows a considerable increase in soil clays in the B horizon which could be expected in a decent weathering profile. Thus it is apparent that significant weathering which produces structured soil profiles in this case is not initiated before a timespan of some tens of thousands of years in temperate climates. However, the formation of clay minerals in rocks exposed to more severe climatic conditions will occur in significantly shorter timespans, as can be noted on any stone building of 100 or more years of age.

The clay forming processes will be slower in colder or arid climates, and will be more rapid in warm climates and those with high rainfall.

Compositional control

The importance of the chemical composition of the source rock is very great in determining the species of clay mineral which will form in stages of intermediate reaction advancement. Certain clays are specific to certain types of source rock or mineral whereas others are found in many types of chemical settings. Figure 4.11 shows the relations between the major magmatic rock type, its major mineral types and the general types of clay which form under conditions of moderate weathering intensity or reaction advancement. Plagioclase, which is generally very unstable, gives rise to kaolinite with loss of Ca, Na and Si in most rock types. Ferromagnesian minerals are more varied in their reaction products which are to a large extent determined by the chemistry of the whole rock. They form first trioctahedral ferrous clays and then dioctahedral ferric clays. Potassium feldspar generally is slow to react and produces either illite (mica) or a form of smectite depending upon its immediate chemical environment or the stage of weathering in which it involved.

Summary

In the structure of the soil profiles, it is important to note the content of the clay minerals with respect to the position in the profile. In undisturbed profiles the greatest concentration of the clays is just below the surface and in the B horizon. Here both chemical alteration and physical transportation increase the clay content. Also in the B horizon one finds that there is a tendency to homogenize the clay mineral species through an interaction of the different clay species present to form a new single mineral. However, this tendency is most often not dominant and one still finds a great variety of clay mineral species in the uppermost horizons of weathering profiles.

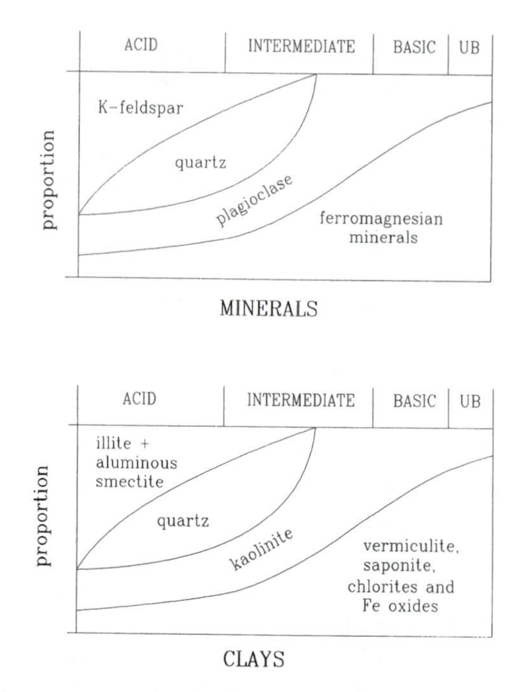

Fig. 4.11 Relations of magmatic rock type (bulk composition variable) and the clay minerals produced from it under temperate weathering conditions. The major magmatic minerals determine, to a large extent, the clays formed. UB = ultrabasic rocks.

Climate modifies the relative importance of the different parts of the profile and dictates the total depth to which clay-forming processes have proceeded. The same is true for the time over which the processes have been allowed to occur. The importance of the slope on which weathering acts also determines the depth of weathering reaction and, to a certain extent, the type of clay minerals formed in the profile.

Therefore, the effects of climate and age of weathering produce similar sequences of weathering profiles. The variables of slope, rainfall and duration of weathering are indicated as they are related to the predominant type of clay mineral formed in surface alteration.

4.1.6 Types of weathering profile in macrocrystalline rocks: examples

Given the importance of bulk source rock chemistry, we will look in detail at weathering profiles of several rock types subaerially weathered in the same temperate climate. We can then compare the clays formed in the different alteration horizons as a function of the rock upon which the soil is based. In these examples, the initial rock type is one with large crystals where it is easy to identify the clay phase produced in its specific chemical environment.

In order to compare the effect of rock type on the clay produced, it is useful to observe alteration of crystalline rock found under the same climate. Examples have been taken from studies in the Atlantic region of western France where the climate is temperate and where rainfall is moderately plentiful. Unfortunately, in most of the examples cited here, the A horizon is probably missing, having been eroded or masked by glacial deposits or outwash action. Nevertheless, most of the description is valid for general purposes.

Granite

Granites weathered under a temperate climate show changes typical of many other types of aluminous, pelitic rocks. Initially, it is important to note that the rock contains phyllosilicates which become clay minerals in a stepwise manner which is not typical of the other minerals present. Muscovite and biotite are phases which are closer in structure and which have a smaller difference in thermodynamic stability from the clays stable under the conditions of weathering. The granite which contains both micas will be discussed as a weathering type.

The weathering profile which includes the structure of the altered rock and the clays produced is given in Fig. 4.12. The profile is divided into four essential parts: rock, saprock, saprolite, and soil. The soil profile nomenclature indicates the R, C, B and A horizons. The representation of the clay content shows that there is a steady increase in clay content as one reaches higher levels in the profile. The reactions are listed in Table 4.1.

The initial magmatic mineral assemblage of plagioclase, potassium

Table 4.1 Clay mineral reactions in granites

Saprock (C)
Plagioclase → kaolinite
Biotite → kaolinite + smectite/mica interstratified
or
Biotite → kaolinite + trioctahedral vermiculite
Saprolite (B)
Muscovite → smectite or dioctahedral vermiculite
Orthoclase + muscovite → illite
Soil (A)
Clays → vermiculite

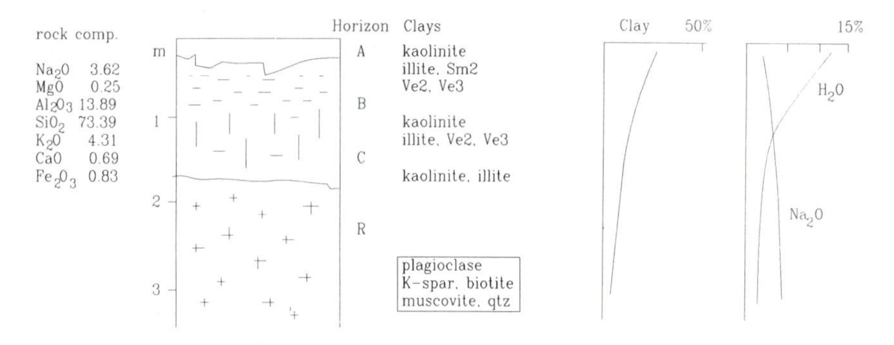

Fig. 4.12 Typical alteration profile of a granite under temperate conditions. Alteration horizons and clay assemblages for each horizon and initial mineral assemblages are shown. The variation of the clay content and that of certain key elements are shown as a function of depth and horizon. The initial bulk composition of the rock is also given. Ve2 and Ve3 = di- and trioctahedral vermiculites; Sm2 = dioctahedral smectite.

feldspar, biotite, muscovite and quartz is transformed, depending upon the local water/rock ratio, into illite and kaolinite minerals in the C horizon. Above this, kaolinite and illite minerals are found with dioctahedral vermiculites (a kind of smectite with non-exchangeable interlayer ions present) and an interlayered vermiculite/illite mineral. In the soil horizon one can find illite, kaolinite, illite/vermiculite minerals, vermiculite and, as a new phase, smectite.

The minerals which successively become unstable are plagioclase, biotite, muscovite and to a certain extent, potassium feldspar. The soil which is the uppermost horizon contains the greatest variety of clay minerals.

The general changes in mineralogy are reflected in the chemistry of the various horizons. H_2O content can be used as a measure of the reaction progress in the hydrolysis of the initial mineral assemblage and hence the clay content. There is a good correlation between the clay content and the H_2O content of the samples. The only other element which changes in concentration to an appreciable degree is sodium. This change is the reciprocal of the hydration of the rock and reflects the dissolution of plagioclase to form kaolinite. Calcium is retained on the smectite-vermiculite minerals. One can surmise that little great chemical displacement takes place upon the argillization (increase of clay content) of the granite under weathering conditions. The changes effected by weathering mainly involve changes in the mineralogy. Basically the pelitic composition is stable under conditions near the surface due to the low mobility of the Si and Al ions present in the initial rock under the chemical potentials operating.

The initial clay assemblage, illite and kaolinite, in the saprock horizon reflects the two extremes of mineral interaction in this horizon. The illite is due to interactions between potassium feldspar and mica, a high-potassium environment. The kaolinite, on the other hand, is due largely to the early destabilization of plagioclase in the zones of fracture and hence high water/rock ratio. In the B horizon the biotites are massively destabilized and one finds an interlayered trioctahedral mica/vermiculite mineral present. Also the muscovites become destabilized internally, forming dioctahedral vermiculites. The increasing instability of the micaceous, phyllosilicate phase enriches the clay assemblage. In the soil horizon, where the clays become an important component of the mineral assemblage, a new mineral, smectite, is formed by a re-equilibrating mechanism through the intimate mixing of the clay minerals in the solifluxed, unstructured zone of the profile. This process is only partial and major inherited phases from the lower levels of alteration are still well represented in the soil mineral assemblage.

The profile shows an increase in clays as one moves upwards in the alteration sequence; kaolinite is more abundant, illite and dioctahedral vermiculite-type minerals are present. Clay mineral heterogeneity is greater with increased effect of weathering.

Basic rocks

Three examples of amphibolite-bearing rocks and a basalt are given here, all weathered under Atlantic, temperature climatic conditions. These could be replaced by examples of pyroxene-bearing rocks, such as gabbros, with little difference in the clay mineral assemblage produced by weathering. The examples used show the influence of the bulk composition of the rock on the clays formed despite an almost identical mineralogy of the initial, high-temperature phases.

Low-alumina amphibolite

The low-alumina amphibolite weathering profile (16.7% w/w Al_2O_3) is developed on a gabbroic amphibolite in western France (Fig. 4.13 and Table 4.2).

Saprock (C) Unaltered rock is found below a depth of 3 m, saprock extends to 1.2 m and the saprolite extends to about 20 cm below the soil horizon. The initial mineral assemblage is plagioclase, amphibole and minor clinopyroxene. Ferric beidellite, dioctahedral vermiculite and nontronite are found, as is zeolite. The dioctahedral vermiculite, derived from a dioctahedral smectite by fixation of non-exchangeable hydroxy complexes in the interlayer ion site, forms in or pseudomorphs the plagioclase

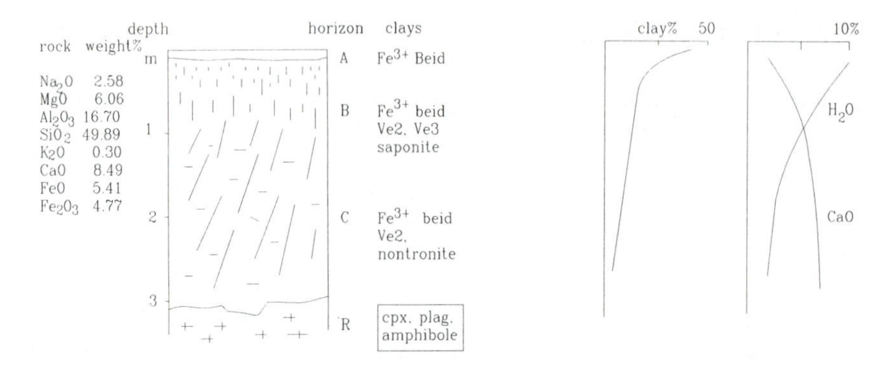

Fig. 4.13 Typical alteration profile of a gabbro under temperate climate alteration conditions. Fe^{3+} beid = iron beidellite, Ve2, Ve3 = di- and trioctahedral vermiculite weathering clays. Initial magmatic rock minerals are cpx = clinopyroxene, plag = plagioclase, and amphibole.

Table 4.2 Clay mineral reactions in low-alumina amphibolite

Saprock (C)
Amphibole → Fe^{3+} smectite (nontronite) + talc
Plagioclase → zeolites + dioctahedral vermiculite (smectite)
Amphibole + plagioclase → ferric aluminous smectite (beidellite)
Saprolite (B)
Clays (ferric and aluminous smectites, vermiculites) → trioctahedral vermiculite + ferric beidellite
Soil (A)
Di- and trioctahedral vermiculite → ferric beidellite

crystals for the most part and the ferric nontronite pseudomorphs the amphiboles. Some talc is also found in the amphiboles. The chemistry of the initial mineral dictates the composition of the clay which forms. However, the clay species produced belong to the same mineral group, smectite.

Reactional zones between amphibole and plagioclase can form ferric aluminous smectite (beidellite). Amorphous materials (gels) were identified by electron microprobe in several sites in the mineral assemblage.

Saprolite (B) The clay mineral assemblage is increased in number of species present with the addition of a trioctahedral vermiculite and saponite (trioctahedral smectite) in the B horizon. The mineralogy of the upper region shows the presence of ferric beidellite and iron oxides illutriated from the A horizon. This change in the mineralogy of the clays is very

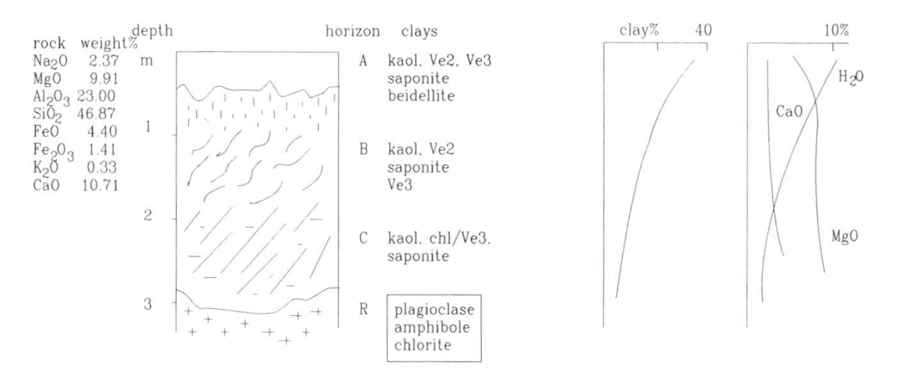

Fig. 4.14 Alteration profile of an aluminous amphibolite under temperate alteration conditions. Kaol = kaolinite; Ve2, Ve3 = di- and trioctahedral vermiculite; chl/Ve3 = chlorite-vermiculite mixed-layer mineral. Plagioclase, amphibole and chlorite are the initial minerals in the unaltered rock.

important in that the trioctahedral minerals essentially disappear and are replaced by ferric smectites. The clays are accompanied by an increased amount of iron oxide.

Soil (A) In the uppermost horizon, there is a great loss of magnesium and a transfer of iron from some of the silicate phases to an oxide. A portion of the ferric iron forms nontronite (ferric smectite) which is the major clay in the A horizon. We can characterize the A horizon of weathering of this basic rock as one dominated by the change in oxidation state of iron which destabilizes many of the weathering clay minerals and which transfers a singificant portion of the remaining elements from silicate to oxide phases.

The clay content increases dramatically in the uppermost zone whereas a steady change in hydration state and calcium loss is noted in the chemical analyses of the horizons.

Aluminous amphibolite Figure 4.14 shows the weathering of an amphibolite in western central France where the initial rock contains only amphibole and plagioclase and much more alumina, 23% by weight, than the preceding case discussed. The alteration profile of the bedrock starts at about the same depth, 3 m. The saprock–saprolite interface occurs at a depth of around 2 m. The A horizon is deeper, about 40 cm, than in the case of the low-alumina amphibolite.

Saprock (C) In the C horizon one finds kaolinite, trioctahedral smectite and chlorite/vermiculite (mixed-layer minerals). The kaolinite forms in the plagioclase and the trioctahedral minerals are largely found associated with the amphibolites which contain major quantities of Fe and Mg.

Saprolite (B) The B horizon contains kaolinite and trioctahedral smectite as well as di- and trioctahedral vermiculites. The same assemblage is found in the uppermost horizon. The new vermiculites are due to the collapse of the rock structure which induces the mixing of the clays. These minerals can influence the chemical environment in the horizon as a whole and begin to form new phases, vermiculites, from the clay formed at the initial stages of alteration.

A Horizon The A horizon shows essentially the same clay mineral association as that found in the B horizon. In this profile developed on an aluminous amphibolite, kaolinite, an aluminous 1:1 clay mineral phase is present in the lowest region of alteration and remains present throughout the profile. This indicates the importance of the higher alumina content of the initial rock as it affects the clays forming in the different horizons and specific chemical micro-environments. Not only is kaolinite present but the alumina in the rock stabilizes secondary dioctahedral clay minerals and preserves the trioctahedral minerals in the uppermost portions of the alteration profile where oxidation would otherwise change them to ferric, dioctahedral smectites. Mg is lost to a much greater extent.

Micaceous amphibolite

This example is given to illustrate the types of weathering clay produced in a rock of complex mineralogy with phases of greatly differing stability under weathering conditions. The rock is a micaceous amphibolite (glaucophane, garnet, epidote, phengite, actinolite) altered, as before, under Atlantic temperate conditions. The initial minerals are varied, some forming clays readily while others are unaffected by the weathering process (Fig. 4.15 and Table 4.3).

Rock (R) The major phases are amphiboles, sodic (glaucophane) and calcic (actinolite). As in the previous examples, these minerals are readily altered into clay minerals. Garnet is also a highly reactive phase, rich in ferrous iron which is readily oxidized. Aluminous mica is also altered but less readily than the ferromagnesian phases. Epidote, on the other hand, seems to be unaffected by the alteration processes.

Saprock (C) Amphiboles are transformed into saponites (trioctahedral smectites), garnet forms kaolinite and iron oxide phases in internal cracks. Muscovite is reactive at contacts with other minerals, garnet and amphibole, forming kaolinite and oxides.

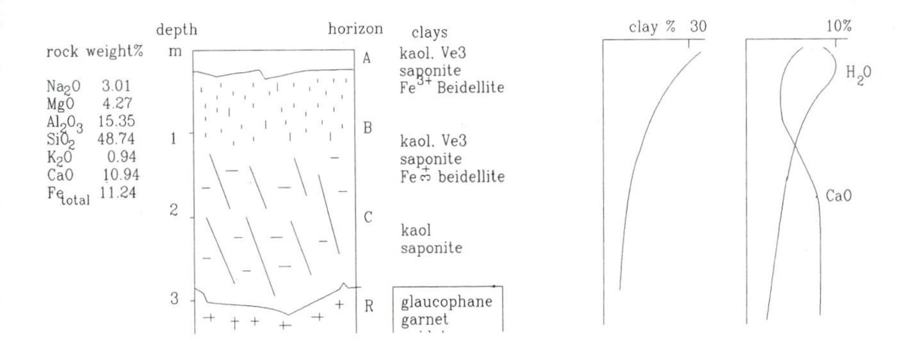

Fig. 4.15 Alteration profile of a glaucophane-garnet amphibolite under temperate alteration conditions. Kaol = kaolinite, Ve3 = trioctahedral vermiculite, Fe^{3+} beidellite, saponite are the weathering clays, and glaucophane, garnet, epidote and muscovite are the initial rock minerals.

Table 4.3 Clays in micaceous amphibolite

Saprock (C)
Glaucophane → Na saponite
Actinolite → low-alumina saponite
Garnet → kaolinite + iron oxides
Phengite + garnet → kaolinite + iron oxides
epidote → unaffected
Saprolite (B)
Glaucophane, garnet, muscovite, epidote as in saprock weathering clays (kaolinite + saponites) → trioctahedral vermiculite

Saprolite (B) In this horizon the amphiboles are intensely altered to saponites, garnet and muscovite form kaolinite and oxides. In the unstructured, upper zone of the B horizon a new vermiculitic mineral is found in the clay matrix. It is the reaction product of the new combination of the clay phases formed through internal transformation of the initial minerals. The new contact between these clays creates a new, trioctahedral mineral with smectitic affinities. In pore linings and along cracks a new, ferric beidellite is present as a new clay mineral.

A Horizon Here the saprolite minerals can be found in various proportions. The trend of mineral change is controlled in this profile by the internal destabilization of the constituent metamorphic mineral assemblage. Only in the upper part of the saprolite horizon can one find new clays forming previously formed clays. The mineral type is trioctahedral, reflecting the importance of the trioctahedral smectites (saponites) which form in lower horizons.

The change in chemistry with alteration and the resultant clay formation shows a marked loss in the calcium content with a general increase in potassium. This latter change is due to a relative increase in the initial mica and a certain retention of potassium by the saponites and vermiculites relative to the other elements. The relative proportions of the other elements remain about constant.

As in most of the cases cited above, the clay content increases in the B horizon to about 25–30% and remains so in the A horizon.

Basalt

Weathering of basalt is difficult to compare to that of crystalline rocks in that the basalts tend to be more difficult to penetrate, resisting locally to form irregular boulders of unweathered material in zones of high clay content. There is a great heterogeneity within a given horizontal plane. This is due to the very compact structure of basalt, frequently containing glassy and microcrystalline material difficult to penetrate. Large fractures or joints provide the major pathways of aqueous access to the mineral core of the rock. These pathways tend to be greatly altered, leaving islands of unaffected rock irregularly scattered through the alteration sequence.

In the C horizon, one finds halloysite ('hydrated' kaolinite), smectite and iron oxides. Ca is lost in the basalt profile as in the amphibolites. It seems that the presence of alumina favours dioctahedral clay minerals which do not contain divalent ions to any great extent. This action then promotes the loss of magnesium to the aqueous altering solution since it does not find a place in the silicates and does not precipitate as an oxide under the conditions of soil clay formation.

Ultrabasic rocks

Ultrabasic rocks are chemically notable in that they contain low alkali, Ca and in the case studied, only 2.82% alumina by weight. The essential components of the rock are SiO_2, FeO and MgO. In such a rock composition the obvious changes at the uppermost level, near the surface, will be the oxidation of iron (Fe^{2+} to Fe^{3+}) and loss of magnesium as in the low-alumina amphibolite. The profiles in ultrabasic rock weathering are shortened in the B and C horizons. In Fig. 4.16 the unaltered rock is found at a depth of 1 m and the A horizon at 30 cm. The mineralogy is very simple, as might be expected from the restricted range of elements and phases present in the initial rock. The original minerals are olivine,

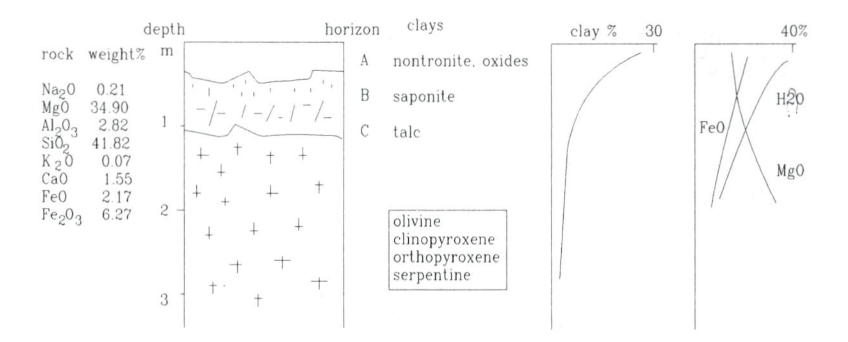

Fig. 4.16 Alteration profile of a serpentinized peridotite under temperate climatic conditions. Nontronite, saponite, talc and oxides are the weathering minerals, while olivine, clinopyroxene, orthopyroxene and serpentine are the initial minerals in the rock.

Table 4.4 Clay mineral formation

Saprock (C)
Pyroxenes → talc
Saprolite (B)
Talc + serpentine → saponite (trioctahedral smectite)
Soil (A)
saponite → ferric smectite (nontronite)

clinopyroxene, orthopyroxene and serpentine. Reactions are listed in Table 4.4.

Saprock (C) The C horizon contains talc as the weathering clay phase, a mineral found only in minor quantities in the amphibolites. The talc is formed essentially at the expense of the pyroxenes.

Saprolite (B) The B horizon contains trioctahedral smectite (saponite). This mineral replaces the talc formed in the C horizon through a combination of serpentine destabilization and the formation of small quantities of an amorphous gel. The trioctahedral mineral is highly magnesian.

Soil (A) The trioctahedral smectite (saponite) is replaced in the uppermost horizon by ferric smectite (nontronite) and iron oxide along with a certain quantity of gel. Again the soil (uppermost A horizon) weathering conditions effect a change of oxidation state in the iron present (silicate phases) destabilizing a trioctahedral smectite with a transfer of a part of this element from silicate to oxide phase. The new mineral is trivalent and dioctahedral. The clay content increases to 25% in this region of the profile. It is also seen that magnesium is lost to a great extent and iron is slightly leached but essentially all oxidized.

The major characteristics of peridotite (ultrabasic rock) weathering are a simplification of the clay species present relative to more aluminous basic rock types such as the amphibolites mentioned above and a shortening of the alteration profile. Also the loss of magnesium is more important than in the other basic rock alteration profiles.

The clay minerals change from trioctahedral forms to ferric dioctahedral minerals. This latter trend is the most important in the weathering of basic rocks and is accentuated in the ultrabasic rock types.

Generalizations on the alteration of crystalline rocks

In the examples given above which indicate the weathering of acid, basic and ultrabasic rocks in a generally similar climatic context, one can compare the trend in the type of minerals and the depth of the different zones of the profiles as a function of the chemical composition of the initial altering rocks. The granite (acid) and amphibolite (basic) rocks have about the same depths of weathering and similar development of the A, B and C horizons. In the basic and ultrabasic rocks 2:1 talc is formed in the lower levels while the uppermost horizon is dominated by the ferric, dioctahedral mineral nontronite. The trioctahedral clay minerals formed in the B and C horizons are destabilized, indicating their inherent instability under conditions of oxidation and high water/rock ratios. By contrast, the more aluminous rocks tend to conserve the early formed minerals and add new phases to the clay mineral assemblage in the upper portions of the weathering profile. However, dioctahedral clays are dominant in these horizons. It is apparent that the trivalent ions are the most stable in earth–atmosphere interface conditions.

This stability produces dioctahedral clay minerals, ferric oxides and gives divalent ions to the altering solutions, such as Ca and Mg. Of course alkalis are lost in all cases, early in the alteration process. As a result, the sediments and the solutions which transport them contain insoluble Si and Fe^{3+}, Al^{3+} minerals and Ca, Mg, K, Na dissolved in the aqueous solutions. These elements will be found in the solid phases only under conditions of concentration of the aqueous phase provoking the precipitation of silicates, salts or under other conditions carbonates.

It is important to note that in weathering profiles developed in temperate climates (a large part of the inhabited earth) smectites, either di- or trioctahedral, are found in the lower portions of the weathering sequence; new clay minerals form from these clays in the middle regions. They are often ferric phases and kaolinite. The latter mineral is common to most of the profiles in varying quantities. Kaolinite is not necessarily a product of intense weathering in temperate climates but a natural consequence of the instability of an individual phase. It occurs with other clays at any level in the profile. For instance, when garnet, a highly aluminous mineral, destabilizes, it forms the aluminous clay kaolinite. The proportion of kaolinite present can be due to increased chemical weathering and high water/rock ratios. In these instances it is found along passages of high water flux such as cracks, fissures and pores.

The overriding effect here is the very local chemical control dictated by the host, altering mineral on the new clay formed at its expense. Thus local chemistry is important as well as the general water/rock ratio.

4.1.7 Mineralogical control: weathering in phyllosilicate-bearing rocks

So far in this chapter we have been concerned with weathering where the initial minerals are not stable at surface conditions (amphiboles, plagioclase, pyroxenes, micas, etc.). In many soil profiles or weathering sequences the initial materials which is affected by the weathering process are phyllosilicates which have formed at low-temperature conditions. Under these circumstances the change in mineralogy due to a change in chemical conditions (water/rock ratio) is much less spectacular (from a clay mineralogist's standpoint) than that observed in crystalline rocks. The mineral transformations are more subtle because the differences in structure are smaller and the differences in chemistry are also less important. For example, a rock containing the low-grade metamorphic assemblage muscovite-chlorite will change to illite-trioctahedral vermiculite or saponite. The basic mineral structures are quite similar, the muscovite–illite change is small and the chlorite–vermiculite change is equally small. The result is that the evolution of the weathering profile is slow, in time and chemical space. The chemical disequilibrium (differences in free energy of the two assemblages) is not great. In these soil profiles, one finds a change in the abundance of clays in the different horizons as weathering progresses but there is relatively little change in chemistry. Typically, in the A horizon, there is a great change in mineralogy where the weathering parameters devitalize the soil clays to form a kaolinite + oxide assemblage. Below this level the changes in clay mineralogy are more qualitative than quantitative.

Glauconite transformation

One example of clay mineral change due to weathering will be given here to demonstrate the types of change which one can expect in clay weathering systems. The initial rock upon which the soil is developed by weathering under Atlantic temperate conditions (western France) is a glauconitic sandstone. Glauconite is a clay formed at the ocean sediment–sea water interface. Therefore its origin is one of low temperatures and it is stable in dilute aqueous solutions. It has a high potassium content and is very iron-rich. Under intense weathering conditions, high rainfall, and in well-drained topographical conditions, glauconite has the tendency to destabilize to a kaolinite + iron oxide assemblage. This reflects the loss of alkali to the solution and not the oxidation of iron because the iron in glauconite is predominantly ferric.

In the weathering profile used here as an example of clay–clay transformations, the formation of kaolinite + oxide occurs after the gradual destabilization of glauconite through a series of mixed-layer smectite-glauconite minerals. Figure 4.17 outlines the profile chemistry and clay content. The B horizon is seen to accumulate the clays and iron oxides, much through the effects of illutriation. The clays produced at the surface, largely kaolinite and iron oxides, are concentrated in this horizon. This is seen in the high H_2O and Fe_2O_3 contents of the horizon. Alumina content increases gradually, showing the presence of kaolinite. Potassium content decreases by about a third in the profile, showing that the potassic ferric smectite phase is still present in the upper regions of the profile.

Electron microprobe analyses have shown the presence of secondary, green potassic minerals of lower potassium content than the glauconites in the bedrock. These new minerals, identified by X-ray diffraction as mixed-layered glauconite-smectites, are the result of lower potassium activity of the percolating solutions. A diagrammatic representation of these relations (decrease of potassium content in the solutions with increased weathering or position in the profile) can be seen in Fig. 4.17. The initial rock contains sand and glauconite, a potassic and iron-rich phase. The lowered potassium activity of the percolating solutions, at higher positions in the profile, is reflected by a gradual change in the clay mineral present.

The initial glauconite is a mineral which is formed through a potassium enrichment of sedimentary materials, either iron oxides plus kaolinite or smectite and kaolinite plus iron oxides. In some instances iron has been introduced into the glauconite via the surrounding solutions. In studies of glauconite formation is has been seen that glauconite forms through the mineralogical change of iron smectite (ferric beidellite or nontronite) into a potassic iron mica. In the weathering process noted here, the potassic mica is gradually changed into a smectite (ferric beidellite).

In the C horizon, plasmic glauconite composition material is found in zones outside the initial glauconite grains which are often slightly transformed. The new plasmic glauconite mineral has a lower potassium content than the initial glauconite and its iron content is lower. X-ray

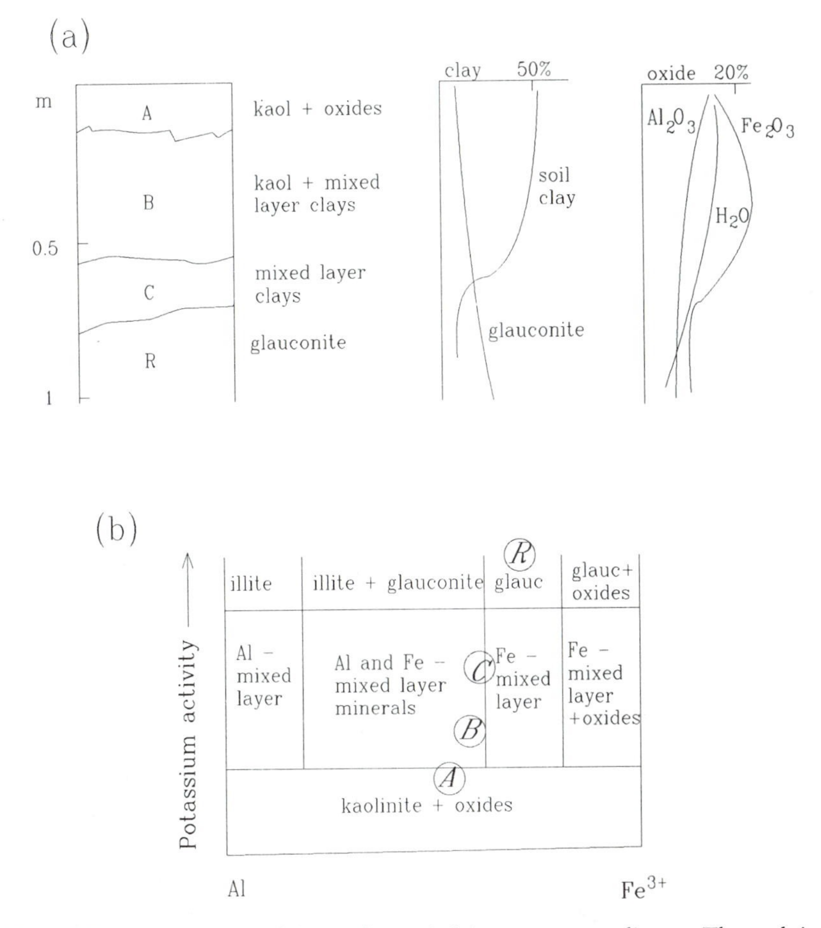

Fig. 4.17 (a) Weathering of clays (glauconite) in a temperate climate. The rock is a limestone containing glauconite. (b) Change in clay mineralogy as a function of the activity of potassium which decreases as one goes upward in the soil profile. The letters A, B, C and R indicate the horizon of weathering in the profile. As weathering increases, the potassium activity in the rock decreases. Therefore the upper portions of the profile are represented in the lower, low-potassium activity, part of the phase diagram.

diffraction of the clays in this horizon shows an increased quantity of mixed layered smectite-mica mineral. Electron microprobe analyses show a gradual change in many points to a new, plasmic mineral which shows a general trend towards the iron-poor composition of ferric beidellite.

In the B horizon, one finds kaolinite as well as the transformed glauconites and the new smectite/glauconite mixed-layer mineral phases. Also there are some zones which contain very aluminous, potassic phases, similar in composition to illite-smectite minerals. There are also many zones of high iron concentration. In the uppermost horizons, and in the fissure zones, kaolinite and iron oxides predominate.

The chemical compositions of the mineral aggregates as a function of the weathering horizon is shown in Fig. 4.17. In this diagram it is possible to trace the mineral assemblages which form due to the increasing water/rock ratio as one moves upward in the weathering sequence of the glauconite-bearing sand. The initial glauconite is changed, gradually to a mixed-layer mica/smectite mineral which has an increasingly lower iron content as the potassium content is lowered. Eventually, in the uppermost zone, kaolinite replaces the ferric beidellite mineral. The iron, liberated as potassium activity no longer permits the stability of a smectite, forms an oxide phase.

In this weathering sequence, an initial clay-bearing rock is altered into a sequence of new clay assemblages which follow a trend of decreasing potassium content and separation of iron from the silicate, clay mineral assemblage.

Soil clay minerals developed on loess and glacial till

In many parts of the world, the first soil used by man for farming purposes was developed upon deposits of fine-grained sediments. These include loess, glacial outwash planes, river outwash plans and deposits of glacial till. All of these sediments are easy to cultivate because they are fine-grained. The particular alteration profile which they develop is dominated by physical transport of clays from the surface to the B horizon. This creates a loose, open texture in the upper portions of the soil zone which are especially well suited to crop growing. The silicate minerals in these sediments are most often dominated by phyllosilicates similar to clay minerals such as muscovite, biotite and chlorite. Many other soils are developed on rocks which contain predominantly phyllosilicates. This the case for most of soils former on sedimentary rocks and a great majority of those formed on metamorphic rocks.

In all of these soils the clays formed during weathering are much less easy to define mineralogically than those formed from wholly unstable precursor minerals such as plagioclase, olivine, pyroxene or amphibole. The reason for this is that the phyllosilicates which are unstable under weathering conditions and which form clays are not all that unstable. They tend to remain present as metastable relicts in the clay structures. The end result is the formation of hybrid-type clays by interlayering of new clay minerals and the old minerals. One typical case is biotite. In many soils it becomes a new mineral with half of its layers in the new chemistry and half with that of the initial or nearly the initial mineral composition. The result in this case is a regularly interstratified mineral of biotite layers and a smectite or vermiculite mineral. The new phase persists in the soil and can be found in sediments and probably in sedimentary rocks.

Illite or muscovite, chlorite and other phyllosilicates have been found to de-stabilize in this step-wise manner, forming complex interlayer minerals. There are probably some three-component minerals formed in soil environments. The end result is the existence of very complex, multiphase assemblages of mixed-layer minerals which are in different stages of change between high-temperature phyllosilicates and those of clay minerals. The vast majority of complex soil clays, such as vermiculites, are due to such processes. These minerals are very difficult to identify because several stages of change can be found in the same soil clay sample. One could have a mica, a half transformed mica-smectite mineral, a three-quarters transformed mineral and a smectite in the same specimen. The XRD, TGA and IR spectra (see Chapter 2) will combine the features of all of these phases which are similar to one another. The resulting determinations combine all of the characteristics of the phases, much to the dismay of the clay mineralogist.

However, since the task of a clay mineralogist is to deal with the cases which are presented to him, and since the soil clays developed on phyllosilicate-bearing rocks are quite abundant, the problem is often present. In such instances, one must rely on the characteristics of the most salient features in the spectra or determinative method. The diagnostics must necessarily be done with caution but the features will be those of end-member types mixed in various proportions. The overall characteristics, such as CEC (see Section 2.7), swelling or other physical properties will be intermediate between the those of the salient end-member minerals, a sort of vague smear of the results. However, these values must be considered as average characteristics of the material which is on its way to a more permanent and stable transformation.

Therefore, in the identification and assessment of many soil clays one cannot use precise mineralogical terms but only qualified terminology which indicates the average mineral characteristics, such as mica-like or smectitic assemblage. The key to an assessment will be in the measurement of the global properties of the material as a function of grain size. Thus the fine-grained material will have certain CEC and swelling properties (expandability or change in average peak position by XRD) and a coarser fraction will have other aggregate properties although the mineralogy will not appear to be greatly different as far as clear mineral species are concerned.

4.1.8 Topographic effect

In each of the examples given above, the topography was similar, the profiles were well drained, that is, the water table was below the alteration horizons but the relief was not great, which allowed a full development of the alteration horizons (that is, water/rock ratios were not great). However, when the topography changes, the depth of alteration changes and the development of the different horizons is affected.

An example of the effect of topography can be seen in three profiles studied at different levels of slope. Again, the examples are taken from an area influenced by an Atlantic, temperate climate (western France). Figure 4.18 shows that the effect of slope is to diminish the depth of alteration as a function of slope. In the steep slope profile, the B (essentially clay-dominated) portion is greatly diminished as well as the C horizon. As slope decreases, the B and then C horizon are seen to become more developed.

The initial rock type, serpentine, can be taken to be similar to that of an ultrabasic rock. However, in the weathering of the serpentinite, one finds that a soil chlorite is present, whereas it is not found in the alteration of ultrabasic rocks or of amphibolites. The change in bulk chemical composition is similar to that of the other basic rocks. There is a pronounced loss of divalent elements, Fe^{2+} and Mg, as weathering is more intense (upper levels in the profile). The uppermost levels are dominated by ferric smectite (nontronite) and iron oxides.

In the B horizon the phases are talc and soil chlorite in the profile developed on the greatest slope, whereas the assemblage is talc, chlorite and ferric smectite in the B horizons developed on the intermediate slopes. The best-developed profile shows only ferric smectite and chlorite in the B horizon. In Fig. 4.18 this change in mineralogy is shown in chemical coordinates. It is evident that the lower the slope (lower water/rock ratios) the more oxidized the iron in the clays and the greater the loss of divalent ions. Also, the greater the slope, the less well developed the profile. The clay horizons are shortened because the chemical effect of weathering is less important. Indeed, the less the slope, the more the chemical effect is evident and the more clay is found and the clay-rich horizons are more developed. More simply stated, clay minerals indicate a change in chemistry.

The same topographic effect is seen in soils developed on gabbroic rocks under slightly more severe climatic conditions (53 cm rainfall per year, towards 20°C average temperature) developed over longer periods of time than those of western France. These are found in Burkina Faso (Fig. 4.19). The alteration profile attains depths greater than 11 m. In the upper portions of the profile one finds a dominance of kaolinite which diminishes in

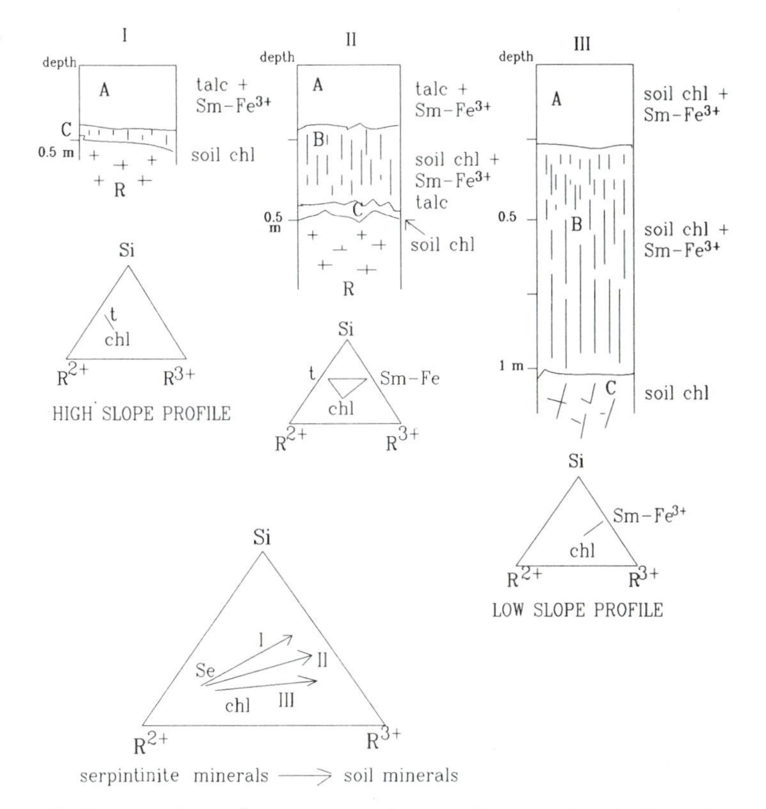

Fig. 4.18 Topographic effect seen in the weathering of a basic rock under temperate climatic conditions. Letters refer to positions in the soil profile. Talc, Fe^{3+} smectite (Sm) and soil chlorite are found as weathering minerals. The rock contains serpentine and Mg chlorite (14 Å). Triangles indicate the chemical evolution of the soil profile as slope changes. I = high slope, II = intermediate slope, III = low slope conditions of weathering. T = talc, chl = chlorite, Sm = smectite. Note that the depth of the weathering sequence changes rapidly as a function of slope. R^{2+} = Mg, Fe^{2+} and R^{3+} = Fe^{3+}, Al.

importance as the profile loses slope. The kaolinite is replaced by ferric smectite (nontronite) in the lower levels of the steeper profiles which becomes dominant throughout the profile in the more poorly drained portions of the slope sequence. Soil or weathering chlorites and talc are found in the C horizons of the weathering sequences, as in the other profiles outlined above.

Here it is seen that kaolinite is typical of regions where the water/rock ratio is high and smectite in regions where the rock has a greater influence on the chemistry of the whole system. Since the profiles are developed on

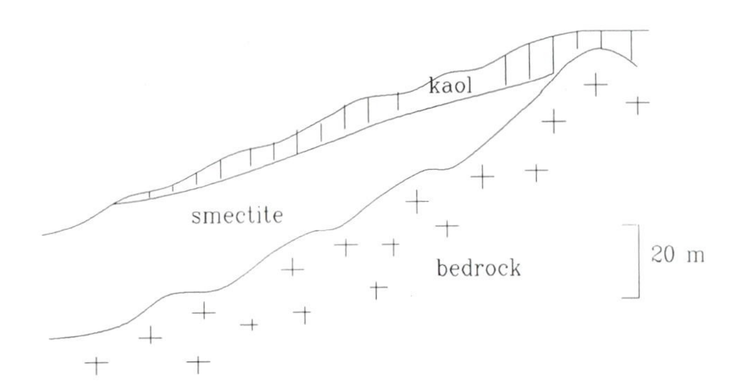

Fig. 4.19 Importance of slope in determining the type of clay mineral formed on a basalt bedrock in a sub-tropical climate. As drainage is less intense, more 2:1 minerals (smectite) are formed.

an ancient heavily eroded continental nucleus and they are all old, that is, the soil or weathering processes have been active for a long time on the same rocks, the profiles of alteration are all more than 10 m deep.

The influence of topography (slope) is very important in determining the type of clay dominant in the upper portions of the weathering profile, ranging from kaolinite in the steeper slopes to smectite in the flatter areas.

4.1.9 Erosion

The development of soils gives rise to clays. This is the most important single source of clay minerals in the geologic cycle. However, soils are formed by the interaction of rainfall with rocks at the surface. The very nature of this interaction is both chemical and mechanical. The chemical interaction produces clays and the mechanical interaction transports clays to another geological environment. Clays are not just born here but they are displaced. The factors affecting their displacement, commonly called **erosion**, are the key to clay transportation at the surface, and an important segment of geological and environmental studies. Man's quest to enlarge his food potential governs the erosion of soils. Clay transport is, then, at present a factor of human activity. The key to the factors of erosion resides in the soils which form the clays.

The study of erosion has hitherto largely ignored the types of clay present in the soil but one can interpret a certain amount of the information given in studies of clay mineral context. Much more study needs to be done in this area.

The initial problem in defining erosion is to establish what process is most important: transport vector (wind, water, ice) or type of erosion (surface rills, gullying, karst or tunnelling at depth). The major, catastrophic types are more a problem for civil engineers who can devise dams, and barriers to stave off the large-scale mechanical transportation of surface material. The surface effects, most common in agricultural areas in developed countries, are very important and have a more direct link with clay mineralogy. Therefore, we will deal here only with the aspects of surface, horizontal erosion instead of deeper, more vertical erosion common to sites of catastrophic erosion. One can note that the erosion of soil recorded in Bavaria has increased by 60% in the last 15 years, mostly without catastrophic erosion. These types of erosion are typical of crop areas and less so of those areas where animals are the major agricultural activity.

One can establish the factors normally considered to be important in the induction of surface erosion of soils:

1. Grain size. It is obvious that small particles will remain in suspension for a longer time than coarse ones. Therefore, the most easily transportable materials will be clays, which are by definition the finest particles present in a soil ($<2\,\mu m$). However, these particles are the most beneficial to plants and hence are vital to agricultural soil.
2. Variability of grain size. The structure of a soil is critical to its vulnerability to erosion. The more heterogeneous a soil (clays, sands and gravels), the more it resists erosion.
3. Humidity and moisture uptake. The more humid a soil, the less it will be affected by sudden rainfall in large quantities (which is the major cause of soil erosion). Therefore, a substantial clay content with significant adsorbed water is a protection against soil erosion.
4. Profile structure. Paradoxically, the less developed a soil profile (more shallow alteration zones) the less vulnerability it will have to erosion. This is simply another expression of the need for a variability of grain size. The well-developed profile shows a thick upper soil zone where clays predominate and this will be subject to clay removal by suspension.

   Thus high clay concentrations in well-developed soil profiles will be unstable when exposed to rainfall. Homogeneous sediments or loess will also be subject to soil clay erosion. The unfortunate thing is that such soils are, and always have been, those which have been used preferentially by agricultural man for about three millennia. However, in western Europe, agricultural man has been using the same soil for those three millennia with relatively little erosion (at least on a catastrophic scale) until recently. Therefore, it seems that new methods of farming are more responsible for surface soil erosion than the fact of farming for a long time on the same land.

Erosion can be checked by respecting the geometric disposition of the soil, that is, slope and proximity to valleys and existing ravines. Erosion can be lessened by using crops which tend to cover the soil with vegetation for the longest time or by leaving the remains of the crop (stubble) in the field for the longest periods to change the structure (homogeneity) of the exposed part of the soil. In fact, most of the realizable active variables which can be used in treating the problem of soil erosion lie in the hands of the person who decides what crop to plant and how to do it. It is difficult to ameliorate the properties of a soil to protect it from erosion by changing an external factor. It is difficult, for example, to change the ratio of clay to sand in a given field without major perturbation and great cost to the owner.

4.1.10 Hydrothermal alteration

The interaction of water and crystalline rock (usually containing few phyllosilicates) at depths great enough to be at temperatures above 250°C, produces clays in localized zones. The type of clay produced depends upon the ratio of rock to water and the anion content of the aqueous solution. The distribution and localization of the clay assemblages in the rocks depends entirely upon the plumbing system (water circulation routes) and the reserves of the aqueous solution. If the escape of the volatile fluids is rapid, it will occur along few and large pathways created by fractures in the rocks. This produces alteration haloes around veins containing clays, carbonates, sulphides or quartz. In some cases valuable elements such as gold, uranium and lead are precipitated with the other alteration phases. In instances where the escape pathways are multiple and the movement of solutions is slow, the amount of rock–water interaction is greater and the alteration haloes are wider, at times invading all of the rock. The extent of the alteration could also be a factor of the length of the alteration process. In any event, in this later case, the amount of alteration, the amount of dissolved material, and the changes in mineralogy are all great. The most striking examples of hydrothermal alteration are the production of pure kaolinite rock or pure silica rock. This can occur on the scale of a cubic kilometre. Both instances are of economic importance.

It is obvious that the process of hydrothermal alteration is an extension of the process of weathering. The reaction products are governed by the same parameters – chemistry (water/rock ratio and rock composition), time and temperature. The types of minerals formed follow the sequence

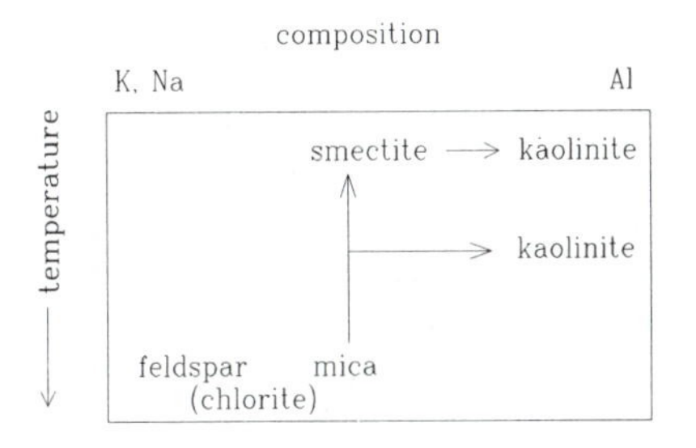

Fig. 4.20 Relationship between temperature and composition in the production of clays in hydrothermal alteration facies. Initial minerals are feldspars, mica and chlorite which form or are transformed into mica (illite), smectite or kaolinite.

noted in the case of diagenetic transformation. The specific assemblages are determined by the rock type and the ratio of rock to solution which reacted to form the clay assemblage.

Figure 4.20 shows the simplified relations of the alteration facies in acidic and intermediate and to a large extent basic composition rocks. At higher temperatures (300–500°C) the clay alteration facies include illite (sericite or mica), potassium feldspar and chlorite. As temperature decreases illite/smectites are found and kaolinite is present. At the lowest temperatures, smectite and especially kaolinite are dominant. Carbonates are typical in the intermediate temperature assemblages.

It seems that the hydrothermal alteration associated with ore deposits is frequently one of hydrogen exchange in the presence of strong anionically controlled solutions. It is a case of hydrogen exchange such as in the reaction

$$HCl + \text{K-silicate} = \text{H-silicate} + KCl$$

This is very different from the situation of rainwater interaction with rocks in weathering where the solution is nearly neutral and anion-free. In the latter case the pH of the solution is affected by the water–rock interaction and the rock has a strong buffering capacity on the reaction. In the former case of ion exchange, the solution pH does not necessarily change and the reaction is carried out depending upon the availability of anions in the solutions. The systematic loss of alkali ions – Ca, Mg and eventually Fe – is striking in the hydrothermal facies. This is less the case in weathering, where alkali and Ca are lost but the minerals tend to conserve Mg and Fe in the silicate or oxide state. This is why kaolinite formed in soils is of little industrial use (because of its iron content) while hydrothermal kaolinite gives white porcelain and other industrial products.

4.1.11 Deep-sea alteration and hydrothermal activity

The interaction of water and rock at depth under the ocean is rather different from that of hydrothermal alteration. The clay mineral assemblage in basic as well as acidic rocks tends to be magnesium- and iron-free in hydrothermal alteration. In the alteration of deep-sea basalts, the clays are in fact iron-rich at very low temperatures. In hydrothermal alteration, the clay assemblages tend to be alkali-free, whereas in deep-sea basalt alteration the clays become more alkali-rich as the phenomenon progresses. Therefore one must assume that the two clay-producing processes occur under different chemical conditions.

In deep sea basalt alteration, the stages of clay mineral formation begin at temperatures below 300°C. In reactions above 300°C the rocks are altered to serpentinites or epidote-chlorite facies rocks. There is a loss of Ca, Na and K. The first clay minerals to form (at 8–280°C) are saponites, often found as fracture wall coatings and vug fillings. This shows a concentration of magnesium and aluminium in the new clays.

The intermediate-temperature clay-forming processes give aluminous and iron-rich beidellitic smectites and celadonites. These clays show a concentration of ferric iron, potassium and aluminium. The celadonite is most striking in that it imparts a green colour to portions of the rocks.

The lowest-temperature clay-forming processes give nontronite, usually of a potassic form. These clays can be found in the 'weathering' of basalts, that is, at the seawater–rock interface which occurs over hundreds of thousands of years. Here the glassy areas are transformed into potassic nontronite. Expulsion of hydrothermal fluids gives the formation of nontronites as the fluids cool to seawater temperatures. These clays are commonly found around black smoker phenomena on the ocean floor. In this facies the clays are almost exclusively formed of ferric iron, silica and potassium.

The general trend of the alteration clay assemblages is shown in Fig. 4.21, where the temperature and composition of the clays are given. A non-clay very often found in the deep-sea basalt alterations are the zeolites. Their chemistry is complex and changes from one site to another. This is undoubtedly the result of difference in chemical conditions but these have not been studied in sufficient detail to give a clear account of the genesis of zeolites in ocean basalts.

4.2 MINERAL STABILITIES

In Section 4.1 it was stated that the origin of some clays was due to the instabilities of other minerals. The clays are the result of incongruent dissolution in aqueous solution. This is largely due to the geological fact

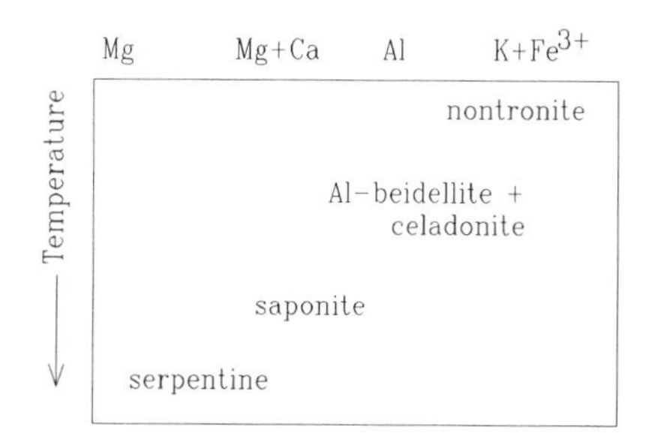

Fig. 4.21 Composition of clays and temperature at which they are formed in deep sea basalt alteration. Lower temperatures favour oxidized, potassic minerals.

that the water (aqueous solution) is more abundant than the silicate solids. This is generally the case in weathering. However, when the aqueous solutions are evaporated, or reach saturation with dissolved species, they can give birth to clays at surface conditions. This is a precipitation reaction. The clays become the stable phase and are no longer the result of incongruent dissolution.

When sedimentary materials are affected by new solution compositions – for example, when terrigenous sediments are deposited in saline solutions (ocean or evaporitic lake deposits) – clay-solution reactions can occur. These are effected by diffusion of elements from the aqueous solutions into the clay mass and structure. There is a transfer of material into and out of the clay mineral as it changes composition. Clays produced in these environments are markers of the sedimentary environment in which they have formed.

As sediments, and the clays they contain, are buried, physical conditions change, notably temperature. The ratio of water to silicate changes also in favour of the silicate. Under these physical and chemical conditions, which are designated as **diagenesis**, the clays readjust to the new constraints by transformation reactions not unlike those known in metamorphism.

These environments indicate conditions where clay mineral stabilities are governing the system rather than the instabilities of other minerals, as was the case in weathering.

4.2.1 Precipitation reactions in soils

In arid and semi-arid steppe lands or high plain soils one often finds carbonate accumulations at or near the surface. These mineral assemblages are due to the climate, which alternates periods of rainfall and drought. The soil accepts the rain and it begins to interact with the silicates in the soil or upper horizons. The successive drought conditions bring the solu-

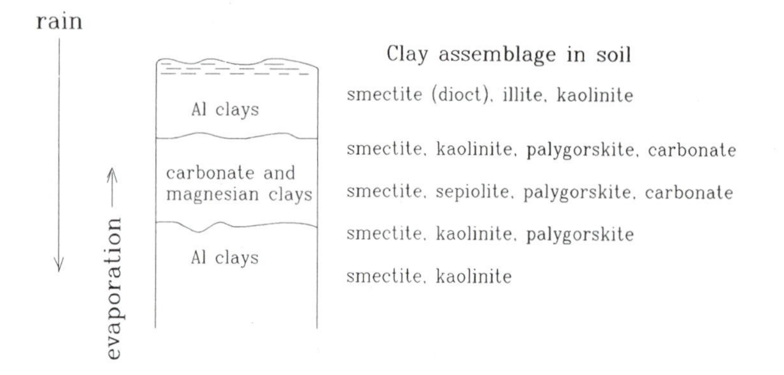

Fig. 4.22 Illustration of the carbonate-bearing soils in semi-arid and arid climates. Smectite dominates the clay assemblages.

tion back towards the surface by capillary action as evaporation dries the upper reaches of the profile. The solutions become more concentrated in the soluble elements which they have acquired in their downward journey through the soil. As the evaporation becomes intense the activity of the ions in solution becomes such that they begin to react with the silicates present. The Mg and Ca ions seem to be the most active in this process. The calcium is combined with the ambient CO_2 to form a carbonate, and the magnesium with the silica and aluminium in the silicates to form the magnesium silicate sepiolite and the magnesium-aluminium silicate palygorskite. Figure 4.22 shows the spatial relations of the old (aluminous) clay minerals and the new carbonate-magnesium clay assemblage. It is interesting to note that there is a gradual decrease in the amount of aluminous minerals present in the total clay fraction due to the combination of aluminium with magnesium to form the aluminous magnesian clay palygorskite. Eventually one finds the purely magnesian clay sepiolite. It would appear that the aluminium is evacuated from the silicates into the lower reaches of the carbonate-bearing sequence and into the lower parts of the profile. The upper levels of the soil profile still contain the initial aluminous minerals which are present in the soil material.

In general, the presence of palygorskite (most frequently) or sepiolite indicate that the soil horizon was formed in arid or semi-arid conditions. These two minerals can be found in shallow sedimentary basins where the conditions of formation are in fact similar.

4.2.2 Sedimentary environment of clay genesis

The point from which clays are removed from their genetic origin in weathering and the next stage of genesis essentially bypasses the stage of

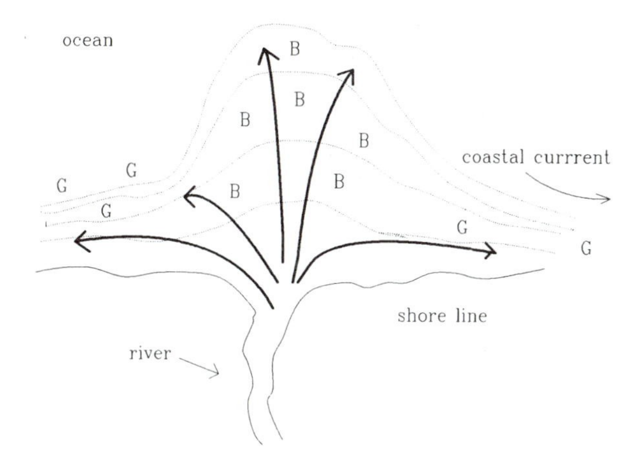

Fig. 4.23 Clay mineral type as a function of near shore, marine deposition facies in shallow water around the mouth of a river. B = berthiérine, G = glauconite. Glauconite is found in zones of lower sedimentation rate.

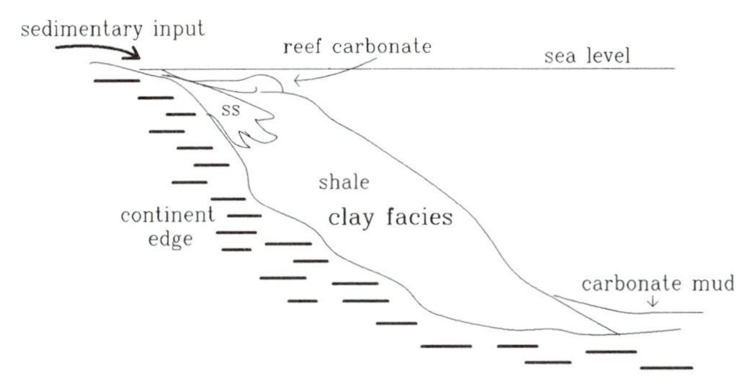

Fig. 4.24 Cross-section of deposition clay facies as a function of the environment (depth) in ocean sedimentation.

fluvial transport. The timespans of this transportational stage are too short to effect clay transformation at surface temperatures.

Transportation of clays in suspension is done by river systems. The deposition zone is spread around the mouth of the rivers and smoothed somewhat by long-shore currents (Fig. 4.23). In cross-section one can see the disposition of the clays compared to the deposits of other sediments such as sands and carbonates in reef situations (Fig. 4.24). The large deposits of clays are found on continental edges situated at the mouths of actively eroding river systems.

The next step in the geological trajectory of clay minerals, the sedimentary environment, is one where certain clay transformations take place and certain clay minerals are uniquely formed. These new clays are rather rare and form a relatively small part of the rock matrix in which they are found; nevertheless they indicate with certainty the sedimentary environment in which they were formed. They are formed by diffusion–transformation mechanisms. The most striking of these minerals is glauconite, and less frequently berthiérine, because they are bright green and even the most unobservant, non-deltonian, geologist cannot miss their presence in a sedimentary rock. The small green oval pellets are an unmistakable indication of shallow-water marine sedimentary environments which have lasted hundreds of thousands to millions of years.

The sepiolite-palygorskite group of minerals are less easily distinguished in hand specimen; they are white or colourless. Nevertheless, they suggest a specific, evaporitic environment which is relatively rare in the normal spectrum of geological phenomena. These clays and others form the clay types whose genesis is that of the sedimentary environment.

The growth of smectites on other smectite substrates is indeed very subtle, yet they form certainly a significant mass of clay material in the initial stages of sedimentation.

The devitrification of volcanic glasses deposited on the ocean bottom after atmospheric transportation into clay-zeolite-silica assemblages is difficult to distinguish from the same effects amplified by burial. If deposition rate is low, many of the smectites will be formed on the ocean bottom–seawater interface.

Although these minerals form rather small quantities of material they are none the less important as markers of the sedimentary environment. The major mechanism of formation is by diffusion–transformation.

Shallow sedimentation

Glauconite

The genesis of these minerals is strictly confined to the sediment–seawater interface or its extension in the very early stages of burial diagenesis. It has been established that glauconite can form only when the sedimentation rate is low, when it remains at this interface as long as a million years. The reason is that the glauconitization process involves the transformation of pre-existing clays, kaolinite, smectite and iron oxides, at temperatures near 4°C and above by means of solid state diffusion or diffusion–crystallization. Diffusion reactions at low temperatures are notoriously slow and the million-year reaction time is not at all excessive for such a process. The precise situation of the origin of these minerals and their potassium-rich composition lends them to use as age-determination minerals. The Sr-Rb, K-Ar and related radiogenic methods have been employed with great success in dating sedimentary rocks containing glauconites. Thus glauconites have become a useful and well-known clay mineral out of all proportion to their relative abundance in nature.

Glauconites are formed at sediment–seawater interfaces where the sedimentation rate is very near zero. The timespan of a million years necessary to complete the process requires a very stable geological environment. This environment is found on stable pericontinental platforms. The depth at which reaction occurs is probably around 100–300 m in most cases. At present glauconites are forming on most of the relatively stable continental coasts. This is especially true of the Atlantic Ocean borders.

The process can be described as follows: clay-rich pellets, such as animal coprolites or shell/test fillings, are subjected to the change in chemical potential of the ocean bottom and probably the Eh potential of the organic residue in the pellet. This destabilizes the clays (mainly kaolinite, smectite and perhaps illite have been identified) incorporating potassium into the new clay matrix and eliminating or accumulating iron as the case demands. The assemblage changes from an aluminous clay with or without iron oxide to glauconite. This can be written schematically as follows:

$$K^+, \text{ kaolinite, smectite, iron oxide} \rightarrow \text{glauconite}$$

$$K^+\ Al_3Si_3 + Fe_2O_3 \rightarrow KFe^{2+}Fe^{3+}Si_4$$

Generally there is a loss of aluminium from the clay mass, and a gain of potassium and some magnesium. A very important step is the dissolution of iron, changing the oxidation state from Fe^{3+} to Fe^{2+}. Intermediate steps in the process, that is, gradual gain in potassium, cause the formation of mixed-layer smectite/mica minerals in the pellet. The mineral layer components are glauconite and iron-rich smectite between the composition of nontronite and montmorillonite. The proportion of glauconite in the mixed-layer mineral increases as the process progresses. Each pelletal concentration of glauconite can evolve on its own. The result of this process is that grains have different compositions in the sediment. Therefore, when XRD spectra are made of a sample of glauconite, the result is a large wide band giving an average intensity maximum for the mixture of the grains present. Each grain is approximately homogeneous in chamical composition (Figure 4.25). The glauconite grains can be segregated into compositional groups by electromagnetic means to give a more accurate estimation of the glauconite compositions in a given sediment.

The geological environment of the glauconite formation tends to be along calm shores with low burial rate, as indicated in Fig. 4.23.

Sedimentary berthiérine

Berthiérine was initially identified as a pelletal sedimentary mineral with an essentially 7 Å structure which contains a large amount of iron, in both

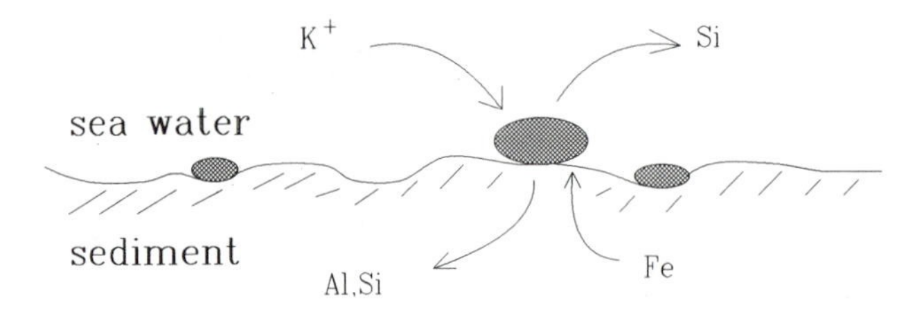

Fig. 4.25 Glauconite formation at the sediment-seawater interface. Glauconites are formed in the shaded pellets found at the sediment-water interface. K and Fe enter the pellet while Si and Al are lost in the glauconitization process.

Fe^{2+} and apparently Fe^{3+} states. The mode of formation is almost identical to that of glauconite. The difference is in the lack of incorporated potassium. It is not known why one incorporates K^+ and the other does not. The most common initial clays in berthiérine formation are kaolinite and large amounts of iron oxide. The determination of the presence of kaolinite as distinct from berthiérine is difficult given the low crystallinity of the phases. Both are 7 Å minerals with very similar (hkl) reflections. Poor crystallinity makes distinctions of (hkl) and (060) bands difficult if not impossible. The mineralogical change is most noted in a change of colour from brown to black and eventually to green. In thin section the berthiérines are brown and the glauconites green. Here it is easy to distinguish the two types of green pellet seen in hand specimen. They can often be found in the same specimen.

The geological environment of berthiérine formation is different from that of glauconite. Glauconites tend to be found along calm, shallow shores of oceanic platforms while berthiérines are found more toward the zones of sediment provenance around the mouths of rivers. This is shown in Fig. 4.23. The occurrence of the two minerals gives some idea of palaeogeography.

Sepiolite–palygorskite

These two minerals are specifically magnesium-rich. When they are found in sedimentary environments, they will have formed in basic evaporitic environments. Closed basins which have sedimentary layers containing these clays show a general progression from the basin edge to the centre of increasing magnesium content of the clay mineral assemblages. The edges are dominated by the mineralogy of the aluminous detrital minerals such as kaolinite and illite. The nearshore sedimentary mineral is palygorskite, the more aluminous of the sepiolite-palygorskite pair. Towards the centre of the basin one finds more and more sepiolite until it becomes the only

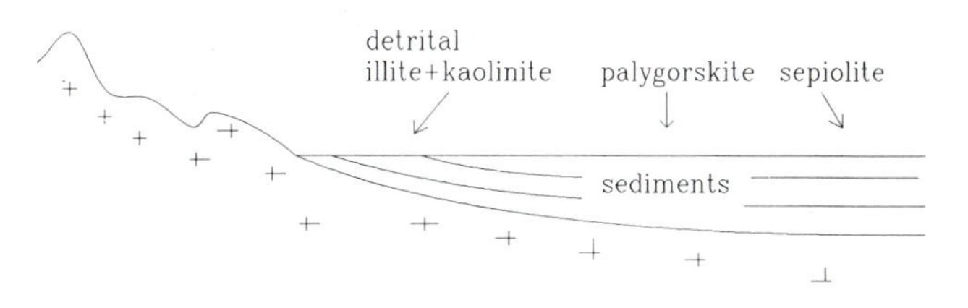

MAGNESIAN EVAPORITIC CLAYS

Fig. 4.26 Cross-section of magnesian clay mineral environments in shallow sedimentation. As one leaves the high depositional sedimentation area, the content of magnesian minerals increases and they become less aluminous (sepiolite).

clay present. Figure 4.26 shows an idealized lake with the sequence of sedimentary minerals present. It is not known if the sedimentary minerals sepiolite and palygorskite are formed at depth in the water or only near the surface when the lake is nearly dry. That is to say, the clays would be produced during periods of drying of the lake basin instead of all at the same time in a basin of water of some depth.

These minerals are often associated with other non-silicates such as carbonates and sulphates. The silicates of authigenic origin which may be present are stevensite and perhaps talc, both of course very magnesian minerals and therefore associated with sepiolite in high-magnesium, low-alumina sediments. The basins in which they are found are often pericontinental or embayed continent edge settings. The reactions of the supersaturated marine solutions with aluminous silicates can be represented as follows:

$$\text{Al-clays (kaolinite, mica)} + Mg^{2+} \rightarrow \text{palygorskite}$$

and

$$\text{Palygorskite} + Mg \rightarrow \text{sepiolite}$$

The introduction of Mg from saline solutions drives the silicate minerals to become more and more magnesian and hence less aluminous. The sodium in the brines does not combine with the silicate minerals because the solutions are basic and alkali minerals (zeolites) are stable in acid environments.

The closed basin origin of sepiolite-palygorskite is thus one of Mg enrichment of the silicate assemblage under basic chemical conditions. These minerals can be accompanied by the non-aluminous smectite stevensite. The increase of the magnesium component is the most striking aspect of this mineral assemblage.

Deep-sea sediments

The deep ocean sediments are slowly accumulating in nature and thus slow mineral transformations can occur at or near sediment–water interfaces. The most interesting, and most difficult to identify, is the growth of aluminous smectite on detrital substrates. The mass of these clays is perhaps not so great but their significance is. This is the demonstration that the smectite minerals are stable in a normal sediment–aqueous environment. Slow growth at low temperatures indicates stability. This is the starting point of the burial diagenesis transformation of smectite to illite. The fully expandable smectites are important phases.

Other smectites, nontronites, are found to form in this environment but they are due to the last stages of hydrothermal alteration of marine basalts and will be considered below in this context.

Another source of clay minerals in the deep ocean environment is that produced during the devitrification of airborne deposited volcanic ashes which cover large portions of the deep Pacific Ocean floor. These materials are, of course, highly unstable and crystallize to form zeolites and clays, mostly aluminous smectites or mixed-layer illites/smectites of low illite content. These minerals persist into the burial environment so that it is not easy to confirm the deep-sea origin. The devitrification process is, of course, a function of time and sediments rapidly buried will retain much of the glassy material until higher-temperature conditions obtain.

The general environments of the different sedimentary deposits which form clays are indicated in Fig. 4.27 according to their depth in the ocean or lake environment.

4.2.3 Burial diagenesis

As the sedimentary basin subsides, and if it has a good supplying river system, it will fill up with sediment. Each successive layer will form small quantities of sedimentary clays with the large body of detrital clays left unchanged. The burial process has the effect of eliminating the free or loosely bound water around the clay particles. Porosity decreases from some 80% to around 20% in the first kilometre of burial. As depth increases and water content decreases, the sediments tend to come into global chemical equilibrium and to form new phases in response to the increasing temperature conditions. Normal thermal gradients in basin sedimentation, either continental or continental edge, are near 30°C km^{-1}. The new phases form according to the laws of kinetics, which gather pace as temperature increases.

Two conflicting effects can be observed as things are buried by sedimentation. The first is that the phases present become less numerous in number of species. The very unstable ones disappear. There is a simplifica-

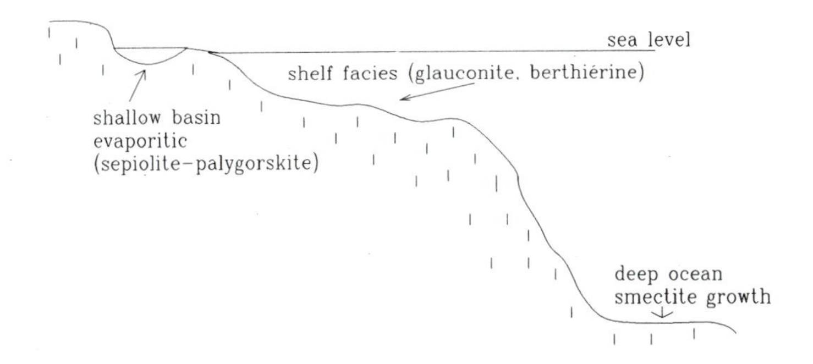

Fig. 4.27 Types of clay minerals found in the different marine sedimentary (depth) environments.

tion in the mineralogy. Also new minerals are formed in a regular manner which changes the mineralogy as a function of the age of the sediment and the thermal conditions which have reigned during its burial. Time and temperature are linked through a constant of reactivity, k, following for a simple case (first-order reaction):

$$dc/dt = k(a - c)$$
$$k = A\exp(-E/RT)$$

where a is the concentration of the initial phase and c that of the new phase, t is time, A is a reaction constant, E is the energy necessary to activate the reaction, R the gas constant, and T temperature (K). Reactions can be considered as being temperature-sensitive, with high E which makes them proceed rapidly at a given temperature, and time-sensitive with low E where they proceed even at low temperatures but over long periods of time. These two effects are shown in Fig. 4.28. The effects of kinetics and greatly enhanced as sediments (clays) are more and more deeply buried.

Diagenetic facies in claystones

In the framework of time and temperature, one can define several typical assemblages of minerals occurring with clay minerals. These will occur as a function of the temperatures they have experienced and the timespans over which they have been subjected to the temperatures.

At very low temperatures and for short periods of burial (near the surface) one will find the full range of soil clays which have been formed during weathering. Here the complex interlayered clays formed by the weathering of phyllosilicates (micas and chlorites) are thoroughly mixed

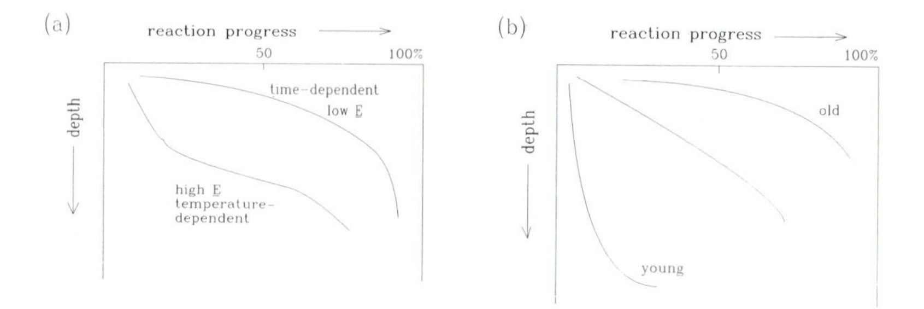

Fig. 4.28 (a) Importance of the kinetic parameters of a reaction concerning the reaction progress profile in a burial sequence. The energy of activation will determine the shape and the depth of the reaction progress curve. (b) Comparison of curves due to different times of reaction in a burial sequence. Reaction progress is greater in older series which give a convex upward curve. In young sediments, the reaction progress curve is concave downwards.

and often make clear clay mineral identification impossible. New clays form as well as zeolites. The clays typical of this facies are sepiolite, palygorskite, aluminous smectites, mixed-layered alteration products and soil vermiculites, glauconite and nontronite, berthiérine and kaolinite. Typical non-clays are alkali zeolites and amorphous silica and cristobalite, ferric oxides.

Intermediate temperatures and timespans give clay assemblages which contain fewer phases. The unstable soil clay minerals in different unstable stages of transformation have disappeared. The magnesian silicates sepiolite and palygorskite also are no longer found. The aluminous smectites begin their transformation into illite and new trioctahedral minerals form. The facies are typically: illite/smectite mixed-layer minerals, 14 Å chlorites, glauconite, kaolinite, mixed-layer chlorite/smectites. Analcite, and quartz are conspicuous non-clays.

The last stage of clay mineral diagenesis is the beginning of metamorphism. The number and variety of phases decreases; some enter into reactions with non-silicates such as carbonates. The major phases are illite/smectite and illite, 14 Å chlorite, corrensite, and kaolinite which can at times disappear. Albite replaces analcite in the zeolite minerals.

Illite-chlorite facies is in fact the beginning of metamorphism. New minerals, often considered to be markers of metamorphism are pyrophyllite and green biotite. However, the clay mineral kaolinite is frequently associated with the illite-chlorite markers of metamorphism. The maximum stability of kaolinite is 270°C according to laboratory experiments. Therefore, the illite-chlorite kaolinite-free assemblage marks the end of clay mineralogy and the beginning of metamorphism.

Illite/smectite mixed-layered clays

The change from smectite to illite follows a gradual change in the average composition of the mixed-layer illite/smectite mineral present in the diagenetic assemblage. This change occurs in several steps. The first is the disordered or random interstratification of illite with smectite. The structure has been given the $R = 0$ nomenclature. The compositional range (proportion of smectite and illite) is from pure smectite to 50% smectite.

At this point the interlayered minerals are found to have another interlayering structure, ordered in a regular interstratification of smectite and illite. When there is more illite than smectite the structure becomes apparently less ordered for obvious numeric reasons. The range in composition of this structure is between 50% smectite (and illite) and less than 10% smectite. The structure is called $R = 1$.

The last stage of the smectite to illite transformation is that of the formation of pure illite. This probably begins early in the sequence of change, occurring with the initial $R = 1$, 50% smectite-illite minerals. Gradual growth of this mineral is seen to be more important in older series of rocks but it seems to always be present. The final stage, when all of the $R = 1$, illitic and slightly smectitic minerals have disappeared, is that of the illite facies.

During the transformation of smectite to illite, significant chlorite is produced either by loss of Fe and Mg from the smectite component occurring during the production of the illite or by combination of kaolinite with iron or magnesium. The change of trioctahedral smectites to chlorite, in sediments of magnesian composition, can produce an intermediate mineral structure, similar to the illite/smectite minerals. This phase is called **corrensite**. It has a chlorite (2:1 + 1) unit and a smectite (trioctahedral) unit in nearly equal proportions. In rocks of special composition, it appears that there can be a gradual change of trioctahedral smectite to corrensite to chlorite as is found in the smectite to illite transition. The interlayering in the chlorite mineral forms seems to be less common than that of smectite and illite.

The overall change in clay mineralogy is seen in a loss of expanding minerals and the formation of a micaceous and 2:1 + 1 mineral.

The different burial diagenetic facies can be shown in time-temperature space as in Figure 4.29. The loss of the siliceous minerals sepiolite-palygorskite and the zeolites is not known as a function of time but an extrapolation has been made from the data for young sediments.

The overall effect of the increase in temperature over different time intervals is a loss of the low-temperature, often expanding minerals in favour of micas and chlorites. The typical metamorphic assemblage in a pelitic rock is mica (or illite), chlorite, alkali feldspar (K and/or Na) and

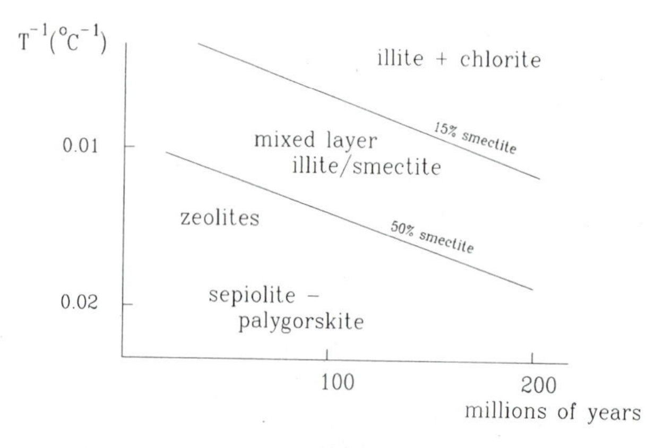

Fig. 4.29 Relations of present-day temperature and age of different I/S mineral compositions coming from burial diagenesis sequences as well as zeolite, sepiolite-palygorskite assemblages. Illite-chlorite assemblage signals the end of the clay mineral facies. Plotting T^{-1} emphasizes the importance of temperature.

quartz. This is indeed a simplified mineralogy compared to the normal clay assemblages in diagenetic rocks containing illite/smectite, illite, berthiérine, kaolinite, quartz and frequently vermiculitic soil clay minerals.

Transfer between rocks: sandstone diagenesis

Sandstones are typified by different amounts of void spaces which are called **pores**; hence sandstones have a certain **porosity**. When the pores are connected they create a permeability. In petroleum geology the quantity of pores and their interconnectivity is very important when hydrocarbons are present. The relation between these two properties determines the

Fig. 4.30 Illustrations of pore wall coating and pore wall fillings in sedimentary sandstones buried to 2 km depth in North Sea sediments. The outside, clear grains are quartz (Qtz) while the area in the centre of the photographs is occupied by sedimentary kaolinite which fills the former pore space (unfilled or filled with oil or water). This kaolinite has obviously grown in the pores given the equidimensional nature of the mineral grains and their disorientation which suggests the absence of transport and physical deposition form suspension. Around the edges of the pores (now kaolinite-filled) one can see a coating of another mineral which, in this instance, is illite/smectite. The mineral grains tend to lie parallel to the pore edge, suggesting mechanical transport and deposition. Thus the pores would have suffered a two-cycle filling, the first by deposition of a small amount of clay from suspension in water and the second by the growth of kaolinite in the pore spaces from Si, Al-rich aqueous solutions.

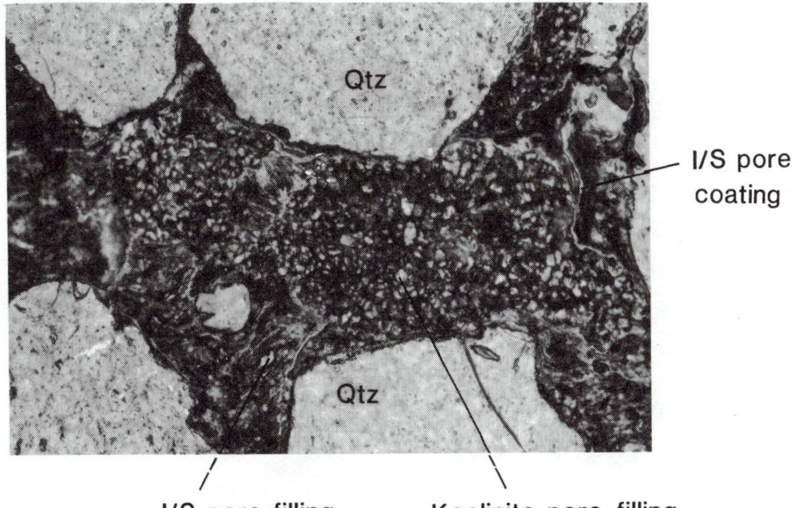
Qtz
I/S pore coating
Qtz
I/S pore filling
Kaolinite pore filling

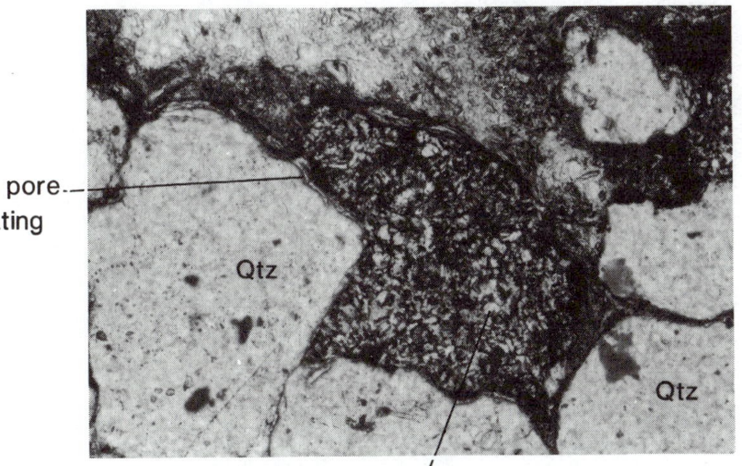
I/S pore coating
Qtz
Qtz
Kaolinite pore filling

4038
10.0U
4048
SNEAP

Fig. 4.31 SEM illustration of two types of clay mineral crystallization in sandstone pores. (a) Kaolinite which fills the pore entirely with its 'books'; (b) illite/smectite crystals growing from quartz grains into the pore space as seaweed grows from rocks into the surrounding seawater.

amount of useful hydrocarbons present. Over the years many devices have been invented to determine the porosity of sandstones and their pore content. These devices are usually lowered into a drill hole in order to assess the productive potential as the well is being drilled. The information is of greatest importance in choosing the methods which will be employed to complete the well and to develop petroleum production. Porosity, permeability and petroleum content are cardinal determinations.

In many sandstone reservoirs, one can find clay particles in the pores and the connecting passages between the pores. These clays have a tendency to change the permeability of the rock. The clays can hinder the passage of fluids. The growth of clays in reservoir pores is part of clay diagenesis or sandstone diagenesis and this aspect of clay mineralogy is very important to the petroleum industry.

Most clays growing in pores show textures which indicate a secondary growth in a cavity. The material which forms the clays must have been transported from outside the pore. In such a case one can suspect a transfer of material between the various layers of sediments and sedimentary rocks during burial and diagenesis. Figure 4.30 shows several clay textural relations found in pores in sandstones.

One typical structure is a sort of clay lining in the pore, where the clays form parallel layers on the pore wall. In this instance one can suspect that the clay was transported physically, that is, as a clay particle, by flowing fluids. These structures are also well known in soils. Such a structure suggests that fluids deposited the clays from suspension on the pore wall.

A second texture is one where clays appear to grow within the pore itself. The most common clay formed in this instance is kaolinite, which often forms vermicular 'books' of clay layer stacks. This secondary growth can be catastrophic for porosity and permeability. The kaolinite growth indicates that Si and Al were introduced into the cavity in solution where the kaolinite grew. There is debate as to how the Si and Al were introduced, through flowing water or by ionic diffusion in more stagnant water. In any event it is clear that the material forming the kaolinite came from outside the sandstone.

Kaolinite is not the only mineral found to grow in pores. Illite/smectite and berthiérine are also commonly found in these sites. One problem has been to distinguish between the quantity of clay and its effect of hindering fluid flow. Blocky kaolinite crystals are easy to identify. They often very

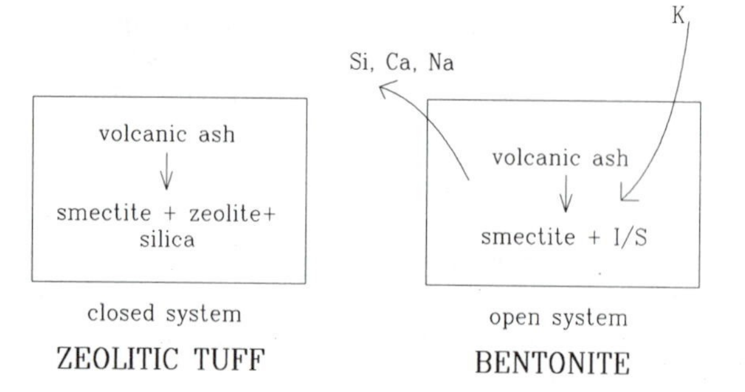

Fig. 4.32 Comparison between the transformation of acidic, glassy volcanic materials into zeolitic tuff in a closed system and into bentonite smectite or I/S minerals in an open system where chamical elements can enter or leave the tuff matrix.

nearly fill up the pore space. However, lath-shaped illite/smectite minerals have a much smaller volume but can obstruct the flow in a pore also. If one does not identify the lath-shaped clays, one will assume that all of the oil in the rock will flow out when it will in fact be obstructed by the clay laths.

Bentonites and tuffs

The devitrification of volcanic ash can give two types of product, one highly zeolitic with varying amounts of clays (tuffs) and the other dominated by clays, notably dioctahedral smectites (bentonites). At present, it is not certain what are the geologic variables which determine the two types of clay-silicate facies. It would appear, intuitively, that the bentonite assemblage, which is often almost monomineral, would indicate an open system of chemical components where a slow diffusion process is made possible by a slow burial rate and hence low compaction rates. The zeolite-dominated facies could be attributed to a more closed system where the recrystallization of the volcanic glass takes place without diffusion exchange.

Figure 4.32 shows the relations of the two geologic systems necessary to produce the clay-zeolite and the bentonite rock clay mineral assemblages. One of the striking characteristics of the bentonite facies is the monomineral assemblage of almost pure aluminous smectite clay. In fact the clays are rarely pure smectite. They are more often a mixed-layer illite/smectite with 70–90% smectite. These clays have been used extensively in industry for their swelling and gel-formation properties.

In many of the bentonite beds one finds a chemical gradation in potassium content which shows changes in the illite content of the I/S mineral. The most potassic (illite-rich) portions are found in contact with the enclosing sedimentary rock layers. This indicates the presence of a diffusion process which gradually transforms the smectite clay into the I/S phase over long periods of time.

Bentonites and alkali zeolite facies tuffs are then formed initially at or near the surface from volcanic ash materials.

5

Uses of clays

Clay minerals have been extracted from their natural environment, as a primary raw material, for most of man's civilized existence. In past centuries clays were used for their properties in producing building materials and ceramics. In this century, clays have become an important part of industrial technology, taking many roles in manufacturing processes, and are major constituents in products such as plastics and foodstuffs. These uses depend upon the special properties of the clay particles. Their chemical properties (internal and external surfaces) are used in many ways, as are their grain size and shape properties.

5.1 INDUSTRIAL USES

5.1.1 Chemical properties

Clay–organic interactions and catalysis

One of the oldest industrial uses of clays is based upon their interactions with organic molecules. Clays have been used as ceramics of different types from the beginning of man's civilization, either as building materials or as the substances to produce culinary ware. These are pre-industrial uses, fundamental but not very sophisticated in the use of the extraordinary properties of clays. Fuller's earth uses the chemical action of the absorbent, interlayer site of smectites or sepiolite-palygorskites to its best advantage. One early problem was to extract the grease, oil and other materials from the wool of sheep in order to be able to use it to make clothing. This was achieved by mixing the wool with clay. Other, more recent uses have been to clarify oils, to clean away unwanted grease and to act as a carrier of insecticides. All these uses take advantage of the highly preferential absorption characteristics of swelling clays and sepiolite-palygorskites. In the twentieth century clays have become very useful in many industries. One major use of clays takes advantage of clay–organic interactions.

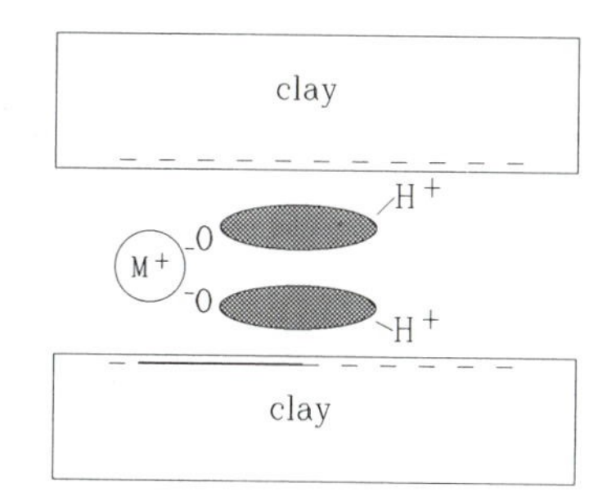

Fig. 5.1 Cation-organic molecule complex organization as absorbed between smectite layers.

What are these characteristics and how do they operate? We have mentioned the tendency of swelling clays to incorporate layers of ethylene glycol between their layers to form a 17 Å complex useful in the identification of their swelling properties. This phenomenon is general, as one can imagine. It is based upon the fact that the basal oxygen layers of the clays which have a charge of less than 0.6 but greater than 0.2 per $O_{10}(OH)_2$ are negatively charged, in an apparently diffuse manner, that is, there is no high charge density at a given point of the surface of anions. Polar molecules are ones which have a free, charged zone at some point on their surface. There is a high degree of chemical interaction and incorporation of such ions between the layers of the smectites and on the exchange sites of sepiolite-palygorskite minerals. The interactions are as complex as is the chemistry of organic molecules. However, one cardinal rule in understanding them is to follow the fate of exposed hydrogens in the organic molecules. As we know, the change of function of these atoms can completely change the properties of the organic molecule.

Absorption of organic molecules

Initially, one must start with the hydrated clay complex which contains interlayer cations surrounded with water molecules. The cations satisfy the negative charge on the clay surface through the hydrogens of the coordinated water molecules. When small organic molecules are placed in an aqueous solution, the water and organic material compete for the interlayer site. As the proportion of the water and organic material is changed in favour of the latter, they are incorporated into the interlayer sites. The initial interlayer reaction, with small organic molecules, is one of cation complexing, with the water playing a secondary role of dissociation. The organic molecules compete with water for association with the interlayer cation. Figure 5.1 shows the type of interaction between cation and organic molecule which can occur when cation complexing occurs.

The effects of clay, organic and water interactions can be followed by

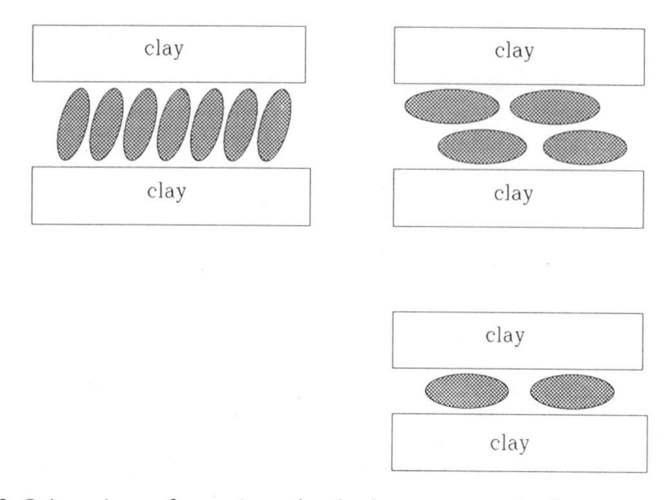

Fig. 5.2 Orientations of organic molecules between smectite layers.

infrared spectra where the bonding energy is seen to shift the interatomic vibrations due to the various chemical bonding interactions.

Larger, or longer, organic molecules can be directly incorporated into the clay interlayer structure without the interlayer cations. The organic molecules are fixed on to sites on the basal oxygen surface of the tetrahedra of the clays by van der Waals or hydrogen bonding. Both types of bond are relatively weak effects which permit a certain mobility of the molecules. Their displacement from the clay site by another species is always possible. In systems where organic materials are abundant in the interlayer surface, they begin to dominate the clay-organic structure. Longer linear, aliphatic molecules (carbon content increases) adopt one of two orientations in absorbed sites, either parallel to the smectite layers or almost perpendicular to them. The structures thus evolved are shown in Fig. 5.2. When the molecules are orientated more or less parallel to the silicate sheet structure, one or several layers of organic material can be present.

In the case of upright orientations, the length of the aliphatic chain determines the interlayer repeat distance determined by X-ray diffraction. This can attain dimensions up to 50 Å, two to three times that of the silicate itself! In fact the layers of organic molecules are not directly upright and, surprisingly, the angle of their orientation varies with the host mineral.

Non-linear molecules, with aromatic components, will assume other orientations, depending upon where the labile hydrogen ions are located in the structure. It has been demonstrated that the interlayer spacing is often slightly smaller than a linear addition of the clay and organic molecule

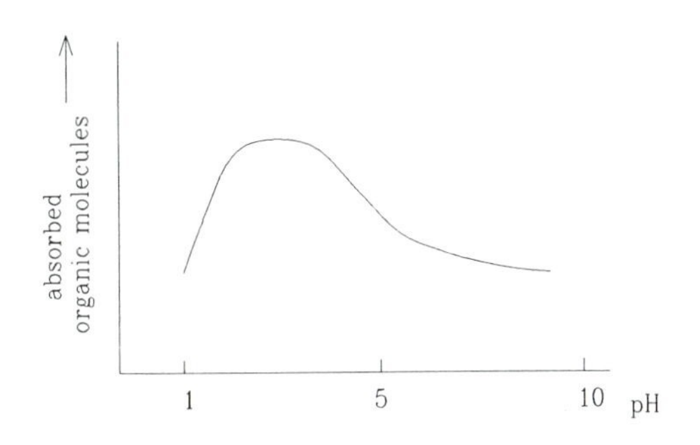

Fig. 5.3 Relation between the absorption of organic molecules and hydrogen ions. The H^+ ion competes with the organic molecule, changing its absorbence on the clay as a function of pH.

dimensions, indicating that the ions in the organic molecules are sited in the hexagonal holes of the surface oxygen structure of the clay layer. This is verified by the shifts in vibrational energy of the appropriate atomic units in the organic infrared spectra. C-H, O-H and C-O bond energies can be affected.

One aspect of clay–organic interactions is the exchange of a hydrogen ion for another cation on the clay surface. If a hydrogen ion is present, this will interact with the organic molecule, often changing its chemical character. This interaction is the introduction to processes of catalysis. Also, if one changes the pH of an aqueous solution containing organic molecules the amount of organic material attracted between the layers of the clay changes due to the competition between hydrogen ions and organic molecules for the exchange sites between the clay layers.

Hydrogen ions can compete with the organic molecules for exchange sites on clays and in using this property it is possible to control the amount of organic material absorbed. Figure 5.3 shows typical relations of pH and amount of organic material adsorbed on a smectite. For pH in range 0–2 and greater than 5, low concentrations of organics are absorbed. By changing the pH one could adsorb and then desorb organic material from clays. This is very useful in various steps in a manufacturing process.

Uses of clays in industry include hydrogenation of different organic molecules. Clarification of solutions through the differential absorbency of heavier molecules is another important use in industry. The impurities which can cause unwanted colouring effects can frequently be corrected through the use of absorbing clays.

Another important use is in fixing organic molecules on chemically inert substrate where the clays can then act as carriers. Pesticides have been frequently conditioned and 'diluted' by fixation on clays. This allows a

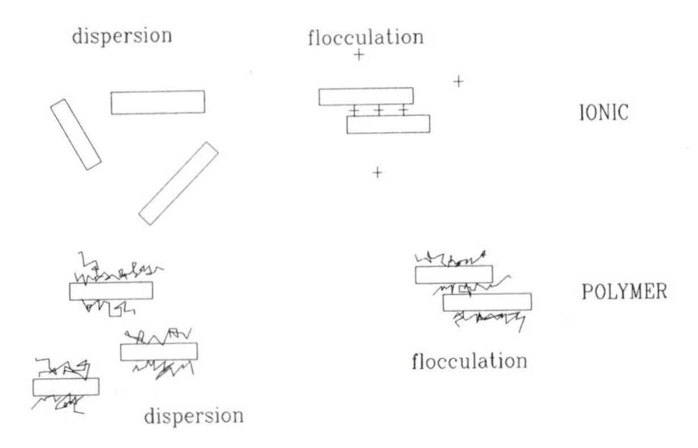

Fig. 5.4 Comparison between the effects of ionic and polymer interaction with charged, clay particles in aqueous suspension. Increasing the concentration of elements in solution tends to flocculate the clay particles.

material to be produced which is easier to handle, being less concentrated than the pure substance. Many other organic products are diluted and conditioned through the use of the absorbent properties of clays. Of course, the major mineral species used are the swelling clays (smectites and vermiculites) and sepiolite-palygorskites.

Polymer adsorption

Smaller organic molecules can be inserted between the swelling layers of clays. This operation is done in a rather geometric manner where the organic molecules form a regular structure in the clay–organic composite. Interlayer spacings follow definite, ordered laws according to the geometry and dimension of the molecules inserted and chemically fixed between the silicate layers. In polymer adsorption, only the clay surface is involved, not the interlayer position. This is due to the irregular forms of the polymers. In order to follow the phenomenon of polymer adsorption, one must go to the theory of clay surfaces and the electrical double layer zone. The non-localized, negative charge on the clay structure is immediately compensated in the zone closest to it by positive ions in aqueous solution. The next zone outward from the clay surface is composed of both positive and negative ions in decreasing abundance. At very low concentrations of electrolytes, the clay particles in aqueous solution are autorepelled by their intrinsic negative surface charge. Additions of electrolyte to the solution attract positive ions to the surface and there is a tendency for two particles to be attracted to the same layer of positive charges, creating a flocculation or coagulation effect.

If polymers are added to an aqueous clay suspension, the same types of process occur. Figure 5.4 shows the effect of addition of polymers to clay aqueous suspensions. As with more simple organic molecules, the initiation of the process is through hydrogen ion exchange or cation interaction with the polymer. Polymer adsorption is conditioned by such effects as pH or exchange cation concentrations. The ideal situation is to coat the clay particles with a polymer, thus producing a stable suspension of clay–polymer in solution.

Clay–polymer adsorption interactions are similar to clay–organic absorptions. However, the range of clay types is greater in that the swelling property is no longer necessary. Kaolinite and other clays can be used in clay–polymer interactions. Kaolinite is very useful because it is very inert chemically and very stable in different pH and oxidation potential environments and so will not interact with the polymer material.

An industrial use of the clay–polymer interaction is that of changing the physical properties of a polymer liquid. Clays tend to give even, more viscous liquids than some pure organic liquids. One such case is in paints which are often extended and conditioned by clay additives. The spreading properties of the paint are enhanced. The dilution property is also important; clays are frequently cheaper than a manufactured organic product.

Catalytic action of clays

A catalytic agent is a substance which speeds up a reaction and is recovered unchanged at the end of the reaction. Natural clays were used during the nineteenth and early twentieth centuries to promote organic reactions, this use reaching its peak in the mid-twentieth century. At present most catalysts are either zeolites or synthetic clays. The rise and fall of natural clays use as catalysts was due to their basic physio-chemical properties and the inherent properties of natural minerals. Clays present two main properties of use in catalysis: a large particle surface area due to their small grain size; and an even larger internal, chemically active surface area due to their absorptive properties. In both instances, the property which is useful is the negative charge on the surface of the clay. This gives the clays the property of a Lewis acid, an electron donor. Such a property is best used in aqueous solution where there is an interaction between the hydrogen and the clay surface, giving an electronically unstable water molecule.

The problems with natural clays are twofold: first, they are non-homogeneous in both quality and quantity (grain size is variable, and the physical properties can vary from one grain to another) and second, they are most stable at conditions near those of their origin – moderate pH and low temperatures. In present-day industrial contexts, temperatures and chemical conditions are more severe than those of the past as processes are made to increase production to higher levels than ever before. Temperatures of hundreds of degrees Celsius and pressures of hundreds of bars are commonplace in factories today, but in the past these conditions could not be maintained for long periods of time. The greatly increased quantities of material to be treated in a given length of time also necessitate catalytic properties of the very highest quality and homogeneity. Thus natural clays have been supplanted in industrial processes by factory products which are more homogeneous and have a higher tolerance of the conditions of the processes.

Nevertheless, the catalytic properties of clays are still important to organic processes which can occur in agricultural and industrial waste contexts. It is therefore useful to know the main attributes of clays as catalysts.

Catalytic properties closely parallel cation exchange properties. It is the charged surface of the clays which is used in catalysis.

External surface properties The most obvious interaction of clays with other molecules is by their crystal surface. Here the properties of swelling clays (internal surface reactions) are not considered; only the chemical activity of the sheet structure on growth surfaces or broken edges of the crystal are considered. The most important effect is that of the activity of aluminium ions in tetrahedral sites. These ions give rise to a negative charge imbalance and hence to an electron donor capacity when the charge is not compensated locally in the crystal. Such a case exists at crystal edges in micas and kaolinites. In previous catalytic processes, the clays would be partially destroyed by acid attack or grinding in order to increase the aluminium ion content at the clay surface and therefore the electron donor effect. The clays can be hydrogen-saturated by cation exchange in acid solution (this is the acidifying process). Such clays are hydrogen donors when presented to organic molecules.

Internal surface catalysis This site is the interlayer exchange ion site of swelling (smectitic) or zeolitic (sepiolite-palygorskite) clays which is specifically due to a charge imbalance in the structure by ionic substitution. Here the surface and charge available is much greater (as seen by exchange capacities of 5–10 compared to 40–120 in swelling clays). The catalytic properties can be further enhanced by providing active metal ions as interlayer cations (Cu, Pt, etc.) which interact in these sites with the absorbed organic molecules. The possibilities of catalysis are therefore much greater and the chemical interactions much more subtle in interlayer site catalysis compared to that of surface catalysis.

Polymerization, **hydrogenation** and **hydrolysis** are all examples of catalytic actions. The individual organic reactions can be more or less speeded up by a given clay depending upon the organic molecules and reactions. That is to say that a given clay will be active for one reaction, and not at all for another of similar nature. Catalysis is always a somewhat

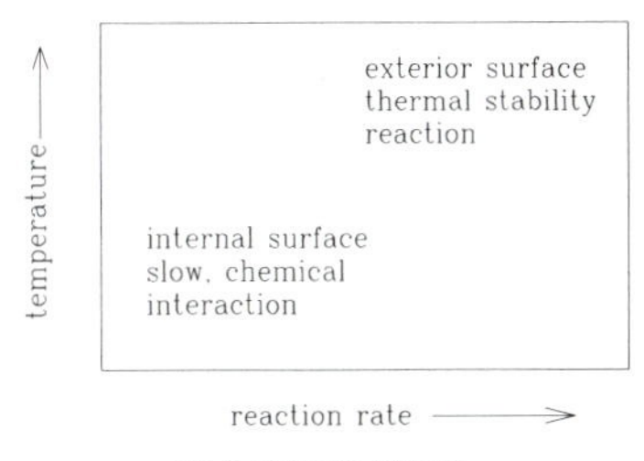

Fig. 5.5 Relation of catalysis type, temperature and reaction rate. The absorption sites between smectite layers is most efficient at low temperatures and low reaction rates. This being the case, new catalysts have often replaced clays in industrial processes.

mysterious process, in that the material does not directly participate in the chemical exchange between the molecules concerned in the process, affecting only the rate at which the exchange occurs.

One of the major uses of clays was in **petroleum cracking** or **depolymerization** of large organic molecules found in natural hydrocarbons. This is, of course, the largest activity of catalysis today in terms of the sheer mass of lighter petroleum products used in the modern world. Since cracking takes place at high temperatures, one would expect that the subtle internal surface properties of clays are not involved much in the process. This is indeed the case, the cracking properties of clays are not related to their exchange capacity or internal surface.

In general one can delimit the catalytic properties of clays as they are related to the conditions under which the catalytic process occurs (Fig. 5.5). The clays with higher thermal stability, such as kaolinite or pyrophyllite, will be used in cracking processes where the temperatures and the rate of reaction are high. The clay surfaces which are readily accessible to the organic molecules are important. The low-temperature reactions such as polymerization or molecular conversion will be more effectively catalysed in slow reactions by internal clay surfaces such as those of smectites or sepiolite-palygorskite. The reactions need time in order to orientate the molecules and introduce the different elements into the structures at the correct junctures.

5.1.2 Physical properties

The particle size and flat grain shape of clays can be useful in conditioning and processing of materials in industry. The small grain size of clays gives more viscous suspensions which can often be used more efficiently in industrial machinery. In such uses, chemical neutrality is considered to be of prime importance. Kaolinite is used primarily in applications where grain size is important.

Grain shape is used to advantage in the paper industry. The clay acts as a filler, taking up the space between the cellulose fibres of the paper matrix. Also, in the presses which process the paper materials into thin sheets, the clays tend to be orientated by the pressure of the rolling process, giving reflective properties to the paper produced. Thus clays (essentially kaolinite) give body to paper by filling interstices, and they also give a shiny surface. Printing inks tend to adhere better to kaolin-treated paper surfaces as well. Thus the major industrial use of kaolinite is in the printing trade. Kaolinites of different grain size distributions and structural states are highly sought after in international markets. Frequently the properties necessary are dictated by the characteristics of the industrial machines used in a given plant. The intrinsic properties of the clay are often secondary.

Both grain size and shape are useful properties in the production of various types of plastic products. The clay charge can dilute and strengthen the final plastic product. The same properties are used in the production of rubber products such as automobile tyres, where kaolinite is frequently employed.

A major use of smectite (bentonite in commercial terms) is in drilling muds. In the process of drilling a deep hole in order to tap petroleum reservoirs, a cooling fluid is needed to allow the drilling bit to operate efficiently. The viscous properties of water–smectite slurries are well suited to this use. Also, if petroleum (or water) is encountered at depth, it is frequently under greater pressure than the hydrostatic head. In such instances the fluid in the drilled hole, in order to be effective, should be more dense than water. The use of bentonite (smectitic) drilling muds thus fulfils two functions at the same time – as a lubricant and as a means of increasing the density of the fluid. However, frequently even the added weight of clay (which has a density of 2.5 compared to 1 for water) is not enough and other materials, such as barite, are added to the slurry suspension to keep the petroleum in its place. In drilling muds the physical properties of smectite, those of making a stable aqueous suspension, are sought after and not the chemical absorbent qualities. The most frequent drilling mud comes from deposits in Wyoming, where thick layers of bentonite occur.

5.2 CERAMICS

A very old use of clays in a transformation process was in ceramic materials. Clays were also used, and indeed still are, by less developed societies as a building material in an essentially untransformed state. Mud-covered houses and pissé walls can be found in many regions of the globe.

The thermal transformation of clays is essentially a characteristic of a developed society, according to archaeological tradition. The cooking and tableware found for various reasons abandoned in the ground has played a great role in the study and classification of past civilizations.

5.2.1 The ceramic transformation process

Thermal transformation uses one of the essential properties of clays, their water content. Initially, the clay is mixed with water to obtain a plastic state. The clay has superficial bound and adsorbed water at this stage. This allows the shaping and moulding of the clay into the form desired by producer and user. The clay is then dried so that it is rigid and self-supporting. The superficial, adsorbed water is lost from the clay in the drying process. Absorbed water is still present in the clay structures. The next step is the firing process, where first the absorbed water is lost, then the crystalline water is lost, and finally, when high temperatures are reached, the clay becomes amorphous and at times recrystallizes. Best-quality ceramics have a mixture of clay or amorphous clay and glass. This amorphous or glassy state makes these ceramics rigid impermeable. Their mechanical properties are increased and the thickness of the ceramic body can be decreased. The ceramic material becomes rigid, and often impermeable. Its surface can be easily decorated.

Surfacing techniques, used to obscure unwanted colours in the clay matrix or to render the material thoroughly impermeable, are many and varied. They can be divided into two types. Slip surfaces are usually superficial coatings of a pure clay, often white in colour. This slip coating hides the colour and sand grains found in the clay body. Slips often render ceramics impermeable. Glazing can be applied to the surface, usually after the clay core has reached a ceramic state. Glazes are glasses which can be either transparent or opaque. They are usually applied to render the surface impermeable and smooth as well as to hide unwanted clay-body colours.

It is essential that the successively added layers adhere to each other after the heating and transformation have been accomplished. This usually involves some fusion interaction between them at their contact points. Too much fusion or interaction will render ceramics shapeless or detract from their surfacing qualities. Therefore, there should be some but not too much interaction between the layers during firing.

The last material is paint, which is applied for decorative effect. Paints are usually oxides, more opaque than the other materials used. If they are to become permanent, they must be applied between ceramic or slip

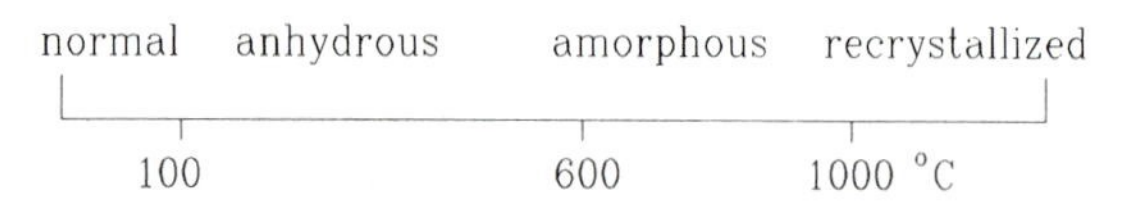

Fig. 5.6 Effects of temperature on the physical state of clay minerals as related to the ceramic processes, producing anhydrous, amorphous and recrystallized forms of the original clays.

coating and the glaze layer. Normally, good paints are embedded into a glassy layer on the ceramic surface.

This brief sketch of the process of manufacture of ceramics has been modified in recent times by the use of different, high-performance materials which replace clays in their traditional roles in pottery manufacture. Different types of vitro-ceram and other products are composed of crystallized oxide and silicate materials which give a very interlocked structure having a thermal expansion of near zero. This differentiates them from most clay-based ceramics which have a non-negligible thermal expansion coefficient. Low thermal expansion diminishes degradation due to repeated heating and cooling, the basis of the cooking process. Further, low or zero thermal expansion allows one to change the thermal regime abruptly – for example, from refrigerator to stove top – without significant change in the properties of the ceramic recipient.

5.2.2 Clay transformations on heating

There are four ranges of temperature which produce characteristic transformations in clay materials (see Fig. 5.6): the free-water dehydration range (50–120°C); the clay stability range (120–600°C); the anhydrous clay range (600–900°C); and the recrystallization range (above 900°C). In the manufacture of ceramics, the 600–1000°C zone is of greatest importance in transforming the dried clay into a new, more rigid substance. In this range the interaction of the clay and non-clay additives occurs to form new minerals or physical states (glass or new crystalline phases). In the clays an important volume change takes place as they recrystallize into other phases, losing their crystalline water above 1000°C. Significant shrinkage occurs as this water is lost (see Fig. 5.7). This loss is important to the ceramic process. When firing brings the material into this thermal region, the shrinkage effect must be modified by the addition of sufficient non-clay materials, called temper or grits. Sands of various types are used normally or, in the case of porcelain, pure quartz and alkali feldspar. The role of temper agents is to diminish the overall volume change occuring when clays lose their crystalline water.

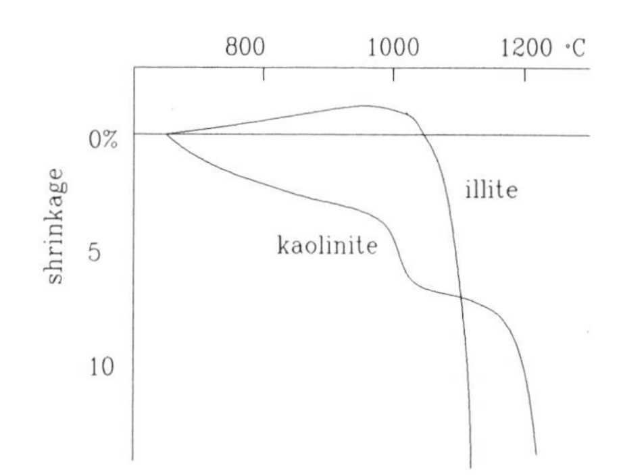

Fig. 5.7 Effect of temperature on the volume of clay minerals. Two contrasting cases are shown: illite, where expansion is followed by contraction; and kaolinite, where any thermal increase effects a diminution of the volume of the clay.

5.2.3 Clay types in ceramics

The proportion and type of clay used in a ceramic product is a function of the cost of the product and its use. Low-cost products will use heterogeneous clays, usually containing oxide and sand impurities. These products will not have strict physical dimension tolerances and will be generally slightly permeable to water. Building bricks are a good example. Significant quantities of sand (temper) are present, about 20–30%. Clays can be smectitic to a certain extent but must contain a fair portion of non-swelling species such as illite, chlorite or kaolinite.

Medium-quality ceramics, such as drainpipes and surface tiles, will have roughly the same clay–temper composition, but the surface will be treated with a material to produce a fused layer to ensure low permeability to aqueous infiltration. Mechanical resistance will be more carefully controlled by a more careful selection of clay materials, mostly illite, kaolinite and chlorites.

Higher-quality ceramics will contain more clay – usually kaolinite, which has low shrinkage – and less temper. However, in high-quality kaolinitic ceramics some material must be added to produce a portion of fused material to hold the ceramic together. This can be calcium oxide and quartz or a feldspar, both of which have a low fusing point. If slip (pure kaolinite) or opaque glaze is used, there can be some impurities in the clay source as its colour after firing is less important.

The highest-quality ceramics (porcelain) require pure kaolinite, quartz sand and alkali feldspar. Firing at high temperature produces a glass in which mullite (alumino-silicate) needles grow into an interlocking grid which gives a high mechanical stability to the product. The clear, white glass gives a translucency to the product. Porcelain requires the highest-quality ingredients and strict controls on the proportions of the components, generally kept secret by the producers.

6

Clays in the environment

6.1 SITUATION OF CLAYS IN SURFACE ENVIRONMENTS

Clays in surface environments are critical to the passage of dissolved species in aqueous solution as they move from one medium to another: air–earth, rainwater–groundwater, sewage–river or ocean, etc. Clays are at the interface of the major surface environments. They are at the surface of the solid earth in its contact with the atmosphere. Any transfer of material between airborne and sediment or rock aqueous systems passes through soils, and hence clays are the critical materials present. The resting place of particulate matter in bodies of water, lakes or oceans, is on clay-dominated sediments. The release or retention of chemical entities in the sediments is governed by clays. Clays are the major inorganic particulate materials in river and groundwaters which will accept or reject ionic or molecular species in aqueous suspensions. The fixation or passage of chemical products will largely be governed by the properties of the clays present. Clays are the substrate of plants and hence will govern the availability of chemical products to plants as they take up nutrients from aqueous solution. They are at the interface between the plant and the mineral world. Clays are present in most aqueous pathways in rocks and hence will be important to the transfer of potential pollutants in problems of nuclear waste disposal (see Section 6.2). Clays are intimately related to the biological processes which are vital to man.

It is evident, then, that we must know how to use our knowledge of clay properties in relation to the problems of chemical transport in aqueous solution. We will therefore follow several examples which give an insight into the importance and function of clay minerals in the environment which man has modified. This concept of modification is very important in understanding the problems concerning the environment as encountered by human endeavour. In natural systems, those where man has not been an active participant, the tendencies toward chemical and biological equilibrium have been manifested for hundreds, thousands and possibly millions of years. For example, the development of a soil profile on a rock body presupposes that the rock is subjected to the chemical agents of the surface atmosphere. The introduction of a rock to a new chemistry is gradual when it is moved upwards in a normal soil and erosion sequence. The subjection of a rock to alteration can take thousands of years or more. The natural erosion processes are normally slow in stable or semi-stable environments, that is, those below the heights of great mountains. However, when agricultural man takes away the normal vegetal cover as he perturbs the surface of the soil-alteration sequence, he changes the rate of reaction between the rocks and the elements of surface alteration. If one considers the reactions which occur at the surface of a rock which has been exposed due to the formation of a new major road, the chemical change is much greater than when a rock is gradually exposed to the effects of climate and rainfall on the floor of a continental deciduous forest.

The main difference between the changes brought about in natural processes and those initiated by man are ones of rate. Man tends to change the landscape at a rate which is not at the same rate as natural processes. For this reason, the study of change brought about by man in natural settings must be taken out of the normal context of natural slow processes and put into one of kinetics. Thus the studies which establish the changes expected in nature must be put in a more rigorous framework where the element of time is considered. The actions of man tend to hasten natural processes.

6.1.1 Chemical pollution problems

In the course of study of clays, it is evident that a knowledge of their properties can help one to understand the interactions of chemicals with the clay mineral interface at the surface of the earth. These properties can be used to assess the amount of interaction and potential hazard in the current problems with which we are faced these days.

In agriculture, the major problems for the last 5000 years or so have been those of erosion and soil clay loss due to intensive destruction of vegetal cover. These problems are essentially those of changes in mechanical properties and stability of rocks and soils suddenly exposed to winter rains. Erosion is still important, but one should be able to take care of it with the old tools of vegetal cover, crop rotation, and other methods of controlled land use.

However, erosion problems might possibly become relatively minor compared to the new threat of chemical contamination. New and intensive uses of pesticides, herbicides and other chemical products in modern agriculture has led to changes in the quality of water sources upon which urban populations depend. Added to this, the industrial chemicals accidentally or carelessly spread on soils and dumping grounds have found their way into the groundwater cycle as a result of industrial usage. The chemical danger is of high potential threat to modern civilization. A clear understanding of the principles of clay–chemical interaction will be a prerequisite for any attempted solution or mastery of such problems.

Two situations are discussed below which outline the principles of clay–chemical interaction which will be useful in such studies. There are, of course, other examples which could be given. One example represents the important aspects of clay–chemical exchange in soil and groundwater aqueous systems. This can serve as a foundation for tackling agricultural and industrial chemical waste problems. The second example is that of radioactive waste disposal which can be used as a second, different type of problem. In the case of agricultural and light industrial chemical contaminations the problem is to restrain spread and to dilute the chemical components over a short timespan, of the order of years. It is basically based upon a 'time will heal' philosophy. In the radioactive waste problem, there is an imperative to isolate the waste from the biosphere for a significant period of time. Here absolute containment is of greatest importance. The two problems are very different, as is the use of clays they entail.

6.1.2 Agricultural and industrial chemicals in the environment

Chemical exchange in soil – groundwater systems

Cases will be dealt with here in an academic fashion, that is to say, without actual field experimental data to substantiate the conclusions and steps in reasoning. The main reason for such an approach is a lack of pertinent data in the scientific literature concerning a given case. The major problem which prohibits a viable analysis in studies of pesticide movement is that the clays and their distribution in the profile have not been studied with care. Thus one frequently has detailed information on the retention and degradation of a chemical species in a natural environment but one does not know why it has been retained nor if the specific properties of clay types present were important in its degradation. One can suspect that clays play a role in chemical degradation – for example, in observing studies of pesticide degradation. Often, very complicated laws of transformation kinetics have been deduced for the loss of chemicals in soils which could be better explained by a conjugated action of microbial activity and clay catalysis or retention.

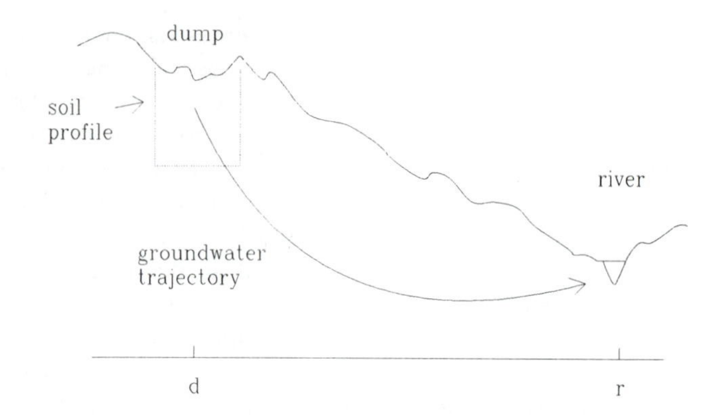

Fig. 6.1 Profile of a pollution site where a chemical has been dumped on a soil profile (dashed line). The curved arrow indicates the normal trajectory of the groundwater as it finds its way to the river.

Cation exchange

The first case considered here is one of a release of a chemical substance into a soil and drainage sequence which allows the water table to be contaminated with the substance in question. The problem is one current in the modern-day world. It is important to deal with the different steps in the process of transmission of the toxic substance in a logical and causal manner. We will use the rudiments of clay mineralogy to demonstrate the expected changes in concentration of the material.

The scenario is classic, one where an excess of a synthetic chemical is introduced upon a soil surface. The information which we would like to obtain is the concentration limits of the chemical which will occur as the solution is moved through the clay substrate of the soil and eventually into the flowing water table, which will be used by the human population at the outlet into a river. Figure 6.1 shows the initial situation. The chemical is introduced on the soil which is linked to a drainage system leading to a river outlet.

The initial stage in the problem is that of the soil profile. Figure 6.2 shows the CEC of the profile which measures at the same time the abundance and the type of clay present in the different portions of the profile. CEC is in general a function of the clay content of the profile at different depths but, of course, it can vary greatly with the type of clay present. In the example given it is evident that there is a concentration of clays in the upper 30 cm depth of the profile. In the simplified case with which we are dealing, the type of clay and its individual exchange capacity are considered to be unimportant, only the total CEC will be taken into account, and all clays are considered to be of the same nature. The CEC is given as a function of depth in Fig. 6.2. The chemical placed in contact

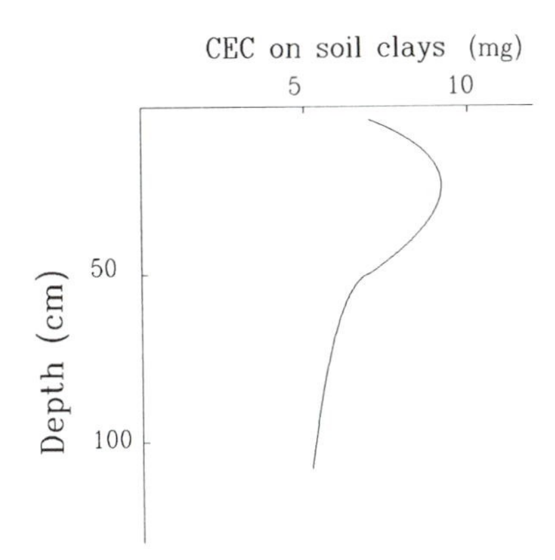

Fig. 6.2 Cation exchange capacity as a function of depth in the soil profile of the dump site. All exchange sites are saturated.

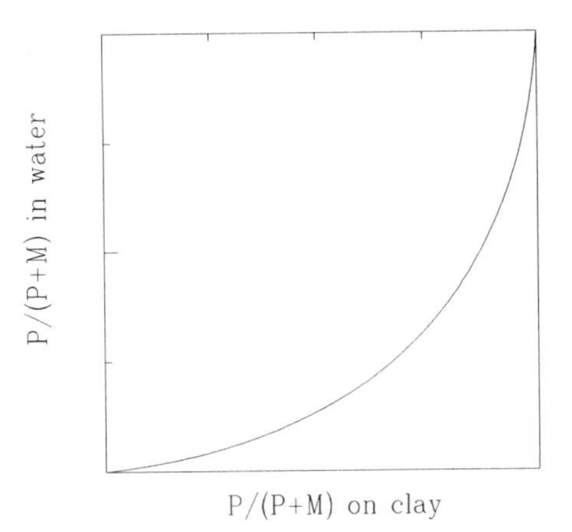

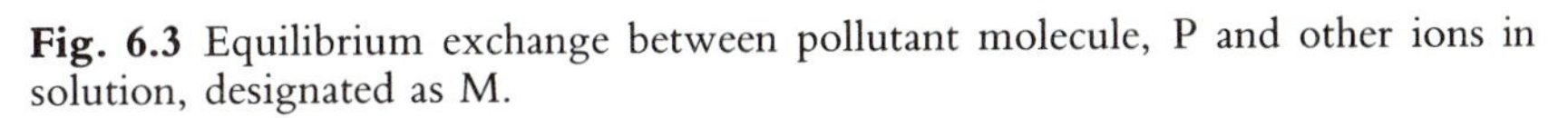

Fig. 6.3 Equilibrium exchange between pollutant molecule, P and other ions in solution, designated as M.

with the soil profile is measured in concentrations of milligrams per litre. For simplicity, we will assume that one litre of solution will come into contact with 100 g of clays in each 10 cm of the soil profile. This gives a slightly two-dimensional aspect to the exercise but it is useful as a simple example.

The exchange isotherm, between the chemical, P, and the other elements in solution will be described as shown in Fig. 6.3, with a *KD* of near 0.5 where

$$KD = (\mathrm{P}_{\mathrm{clay}})(M^{+}_{\mathrm{sol}})/(M^{+}_{\mathrm{clay}})(\mathrm{P}_{\mathrm{sol}})$$

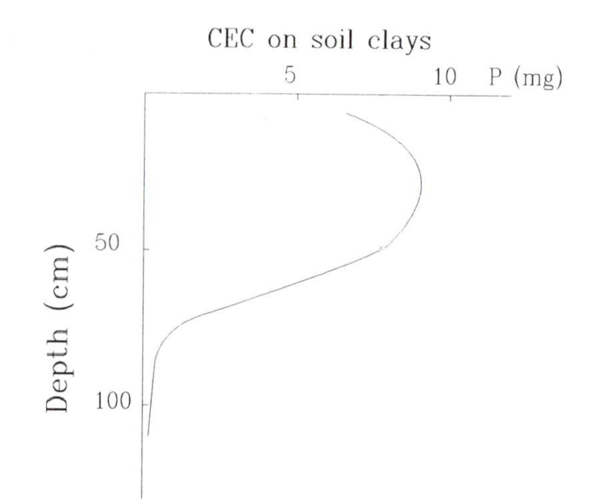

Fig. 6.4 Amount of P (pollutant) exchanged on soil materials as a function of depth when the concentration of P in solution is 45 mg/l in the solutions percolating into the soil profile. All sites on the clays are saturated with P.

It is assumed that the aqueous solution initially contains small amounts of other ionic components (M^+) as it finds its way down the soil profile. The contaminant P is then only one of several components which can be exchanged onto the clay particles. As the concentration of P decreases, the relative concentrations of the other exchangeable ions increase.

We will consider, initially, two possibilities: one where the absolute concentration of P in solution is so great, compared to other ions in solution, that it saturates all of the accepting clay exchange sites in all portions of the profile, and the other where there is not enough pollutant P to saturate all of the sites on the clays.

Saturation

In the first case the clays exchange P to their maximum CEC capacity, which is translated into the concentration of P in the profile (Fig. 6.4). Each litre of solution is successively exchanged with another, deeper, 10 cm of clay as the litre of solution goes down the profile. The content of P in the initial solution must be greater than 80 mg in order to saturate the CEC clay sites to 130 cm depth in the profile. It must in fact be greater than this to counteract the effect of any other ions in solution. At maximum saturation of P there will be at least 10 mg in solution and 10 mg on the clay, and thus the initial solution must at all times have more than 20 mg of P available for exchange. If these conditions are met all of the clays are saturated.

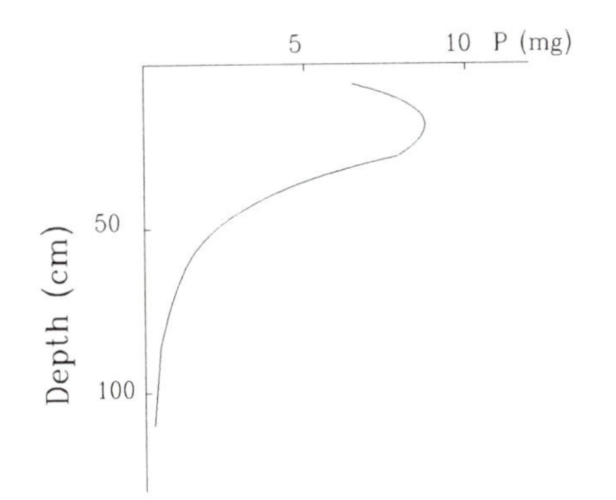

Fig. 6.5 Profile showing the concentration of pollutant P as a function of depth when the initial solutions are undersaturated with respect to the total of the absorption sites of the soil profile, there are less than 45 mg/l of P in the solutions. The initial, surface portions of the profile are saturated and the lower portions undersaturated.

Undersaturation

Figure 6.5 shows the *KD* relations between P on the clays and P in the solutions when the ratio of P to the other ions in solution is such that P does not saturate the clays. Figure 6.5 is important as a means of tracing the concentration of P on the clays when the solution contains less than 79 mg/l. It shows the concentration of P in the soil profile when the initial concentration of P in solution is enough to saturate all of the clay exchange sites in the top of the profile. If one sets the initial concentration at 45 mg of P and 1 mg of M^+ exchangeable ions, the initial part of the profile is similar to that of the case of saturation but only until the maximum concentration of solution plus clay is no longer great enough to saturate the clay. At this point the proportion of P fixed on the clay is a function of the available P related to the M^+ content of the solution. This occurs near a total concentration of 33 mg P at a depth of 30 cm. Below this level the ratio of P on the clays to that in the solution is changed rapidly as total P available decreases below the saturation level. Most of the P available remains on the clay and the solution is rapidly impoverished in the chemical as it moves to greater depth in the profile. The inability of the solution to saturate all of the exchange sites results in a concentration of the chemical in the clay-rich portion near the surface of the profile.

Desorption and migration

We will now look at the case where a soil profile is already saturated with P, to the exclusion of other ions and a new water, containing no P, is introduced into the profile. Figure 6.6 shows the concentrations of P on

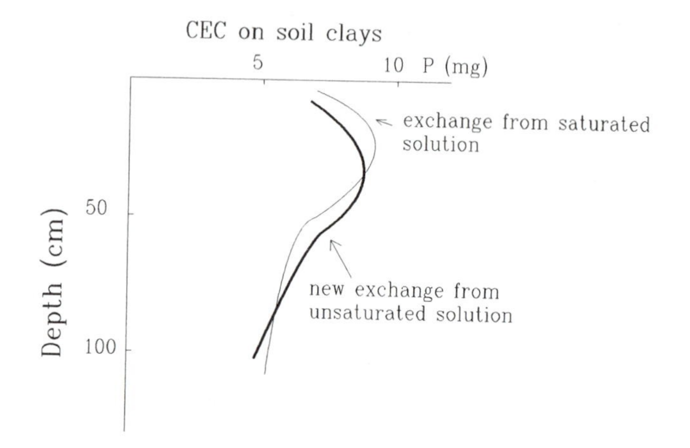

Fig. 6.6 Demonstration of the change in concentration of pollutant P in the soil profile when the profile is initially saturated with P and where new, unpolluted solutions subsequently percolate into the soil profile. The concentration curve of P is displaced lower into the soil profile and pollutant P is distributed deeper towards the water table.

the clays as a function of depth with the introduction of pure water solutions. When undersaturated solutions are put in contact with the clay, there is an exchange between clay and solution (desorption) which gives P to the solution exchanging M^+ ions onto the clay exchange sites. The initial, undersaturated solutions become enriched in the chemical P. As the solutions descend in the profile, they change the concentration of P on the clays, moving it downwards. This is done by introducing P into the solution and moving P further down into the water table. The overall tendency is to flatten the distribution curve in the profile and to spread P into further reaches of the hydraulic system.

Therefore, successive introduction of water into a soil profile moves the exchange chemical into the profile, diluting its concentration. However, the chemical moves further into the profile and towards the end-point of the water table cycle. Figure 6.7 indicates the amount of P in the water solutions flowing through the soil where P was introduced as it moves to a river outlet. This is a typical groundwater trajectory. The initial stage of P concentration in the groundwater is one where the pollutant is concentrated in and near the dump area, time t_1. As rainwater dilutes the groundwater solution, P is exchanged downward into the soil and exchanged with the rainwater. The concentration of P increases in the water as it leaves the clay under low total concentrations of P in the clay-water system. At time t_3, there is enough rainwater to move the pollutant into the riverwater. The concentration is very low compared to that initially

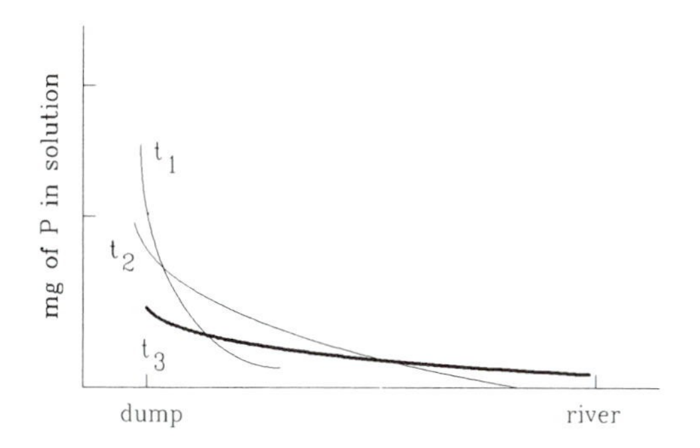

Fig. 6.7 Concentration of P in solution-depth profiles as a function of distance from the river outlet of the drainage profile. The profile is initially saturated with P at high concentrations near the surface. Time t_1 is when the soil is initially saturated, t_2 and t_3 when the concentration of P in solution and on the clays changes as successive waves of unpolluted water pass through the system. The diluting effect produces a decrease in concentration of P in the upper areas of the trajectory and an increase in concentration in the solutions as they move towards the outlet of the river. General dilution brings higher levels of the pollutant into the distant water sources.

found in the dump area. However, there is a strong background of the pollutant in the whole water system. Repeated dumps will of course eventually build up the concentration of P in the water throughout the system.

These interactions between clay and solution assume that there is enough time for the clay and solution to react as the solution descends into the soil profile.

Rapid rainfall input

In the soil–solution interactions thus far described, it is evident that the general result is an enrichment or transfer of the pollutant from the solution to the clay fraction in the soil. There is an immediate concentration of the chemicals on the clay surfaces. This concentration is diluted by successive washings due to renewed rainfall. The concentration of the chemicals near the surface in the initial stages can have a potential positive as well as negative effect.

The normal effect of ion exchange for many artificial chemical products introduced in soil sequences is to concentrate them at or near the surface. Given this concentration of the chemical P at or near the surface, we will now look at the subsequent change in concentration of P between the soil substrate and new, abundant undersaturated solutions in highly non-equilibrium situations.

When sudden, high rainfall occurs during a dry season, for example,

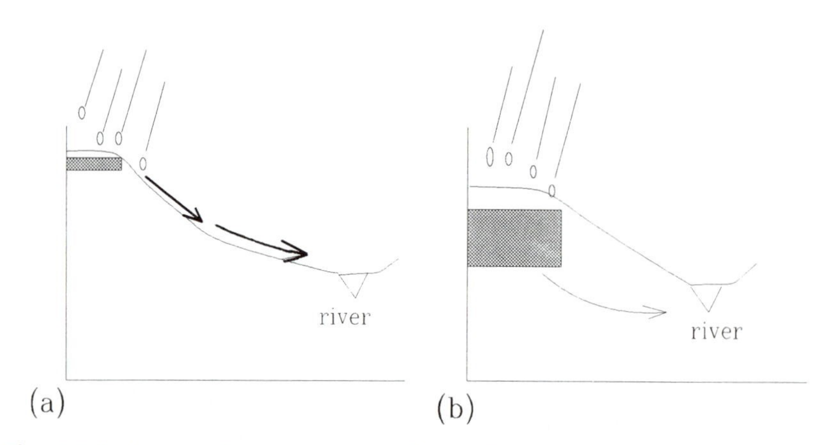

Fig. 6.8 Pathway of displacement of pollutant P in two different rainfall regimes: (a) low rainfall with sudden increase in water content of the system resulting in surface washing and displacement of pollutant P rapidly with particulate matter into the stream area by surface erosion; (b) normal displacement of pollutant into the soil profile and its eventual release into the river system. Scenario (a) presents a great hazard to living matter in a brief, highly irregular period of intense rainfall. The shaded area shows the zone of high pollutant concentration.

two things happen. The first is that much of the rain is moved on the surface by sheet flooding, directly into the stream systems. This runoff typically muddies the streams. This means that the rain has brought a significant portion of the surface clays into the stream system. If our chemical P is fixed onto the clays, they will retain the chemical as non-equilibrium movement occurs. When the clays begin to settle out, in lakes or in larger rivers, the normal exchange of P between solution and clay will begin to occur, releasing it into the water. In this way sudden influxes of agricultural chemicals into stream and rivers can be found to occur during brief periods of heavy rainfall.

The rainwater which does not run off into the streams on the surface will begin its downward journey into the groundwater system. Let us suppose that the influx of solution is very great and that the solution flows through a soil saturated with P into the fracture system of the soil and rock structure without attaining true chemical equilibrium. The solution will pick up, by rapid and partial exchange, a significant quantity of P in the surface area, where the clay and P concentration are the greatest. The rapid flow of the solution tends to keep the P in solution instead of exchanging it onto the clays available at various lower points in the system.

In this case the solution acts as a physical carrier of the contaminant P into the water table instead of exchanging it onto the clays in the upper

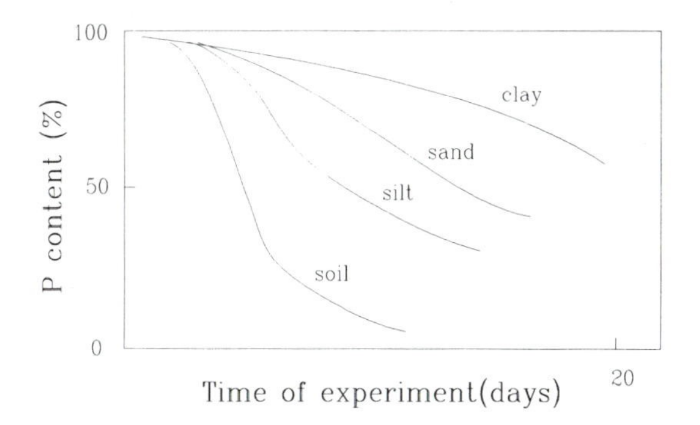

Fig. 6.9 Kinetic factors of destruction of complex organic molecules for different grain sizes in soil profile conditions. It is evident that the smaller the particle, and especially the more absorbent the particles, the longer the retention of organic molecules which are protected from bacterial degradation.

portions of the initial soil profile. Infrequent but regular actions of this sort, which are common in natural cycles in nature, are often the most dangerous. If such a situation occurs once every two years and if the concentration of P is sufficient to kill the fish population in the river below which is linked to the soil profile (Fig. 6.8), the fish population will not recover in the intervening two-year period and the fishing in the river will suffer due to an unusual rainfall every two years in the dry season. In natural systems, the equilibrium of life depends upon equilibrium. The unexpected but foreseeable is enough to upset this equilibrium.

Decay laws

In the several books and numerous articles thus far published on the subject, it is apparent that the destruction or transformation of synthetic chemical products used to destroy plants and insects is a complex matter. The normal laws of chemical kinetics are not followed. Several factors are important in the destruction or permutation of these chemical substances. The importance of microbial activity is cardinal. It has been postulated that any chemical product can be neutralized if the proper bacteria can be found. This supposes a great fund of ambient bacterial types in the soils given to agricultural and other use. However true the idea may prove to be, field tests to date have revealed several factors which are of great importance to people who try to deal with the realities of pollution. It is evident that the absorption of chemical agents by clays takes them outside normal degradation kinetics. The rate of change of a chemical product in a sandy soil is much greater than it is in a clay soil. A shielding effect apparently occurs when the molecules are held between the layers in a clay compared to the state when it is adsorbed on the surfaces of grains and not

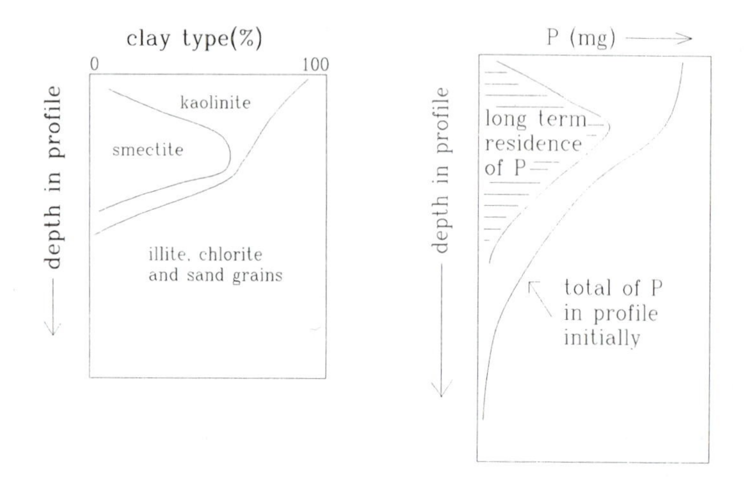

Fig. 6.10 (a) Distribution of clay type and depth in a hypothetical profile. (b) Residence time of a given organic molecule (P) in the profile of (a) compared to the total of P exchanged on the soil materials. It is evident that the type of clay mineral present will determine the resistance of a molecule to organic degradation.

chemically bound. Therefore it is of great importance to know the clay mineralogy of a given soil profile in order to predict how the chemicals will be transformed as a function of time.

If one has the CEC of a soil profile, it is important to know what are the minerals which contribute to this cation retention. If they are smectitic clays the chemicals in solution which exchange will be withheld from immediate bacterial action and gradually released, perhaps during periods of biological inactivity. These relations are shown in Fig. 6.10 where the effect of availability to biological action varies as the clay type differs. The smectites keep the chemical in absorbed interlayer sites which protect it from biological activity. The other clays adsorb the chemical, which is rapidly degraded. The second part of Fig. 6.10 shows the amount of chemical which will remain in the profile after the initial biodegradation period.

6.1.3 Summary

It is apparent that the use of the basic elements of chemical interaction between ionic or polar molecules in solution and clays can be of great help in interpreting or predicting the lifespan and concentration of chemical agents in groundwaters. The exchange and concentration of these chemicals in the upper regions of soils can be a useful point if one is worried about the dispersion of the chemical into the groundwater and eventual water supply of a nearby community. However, if the material remains at the surface it is always a potential hazard liable to be dislodged by sporadic and intense rainfall. The movement of solutions in crevasses and at the surface by runoff can be a great hazard to health in periods of unusual drought.

It would seem that the ideal would be to have the material move slowly into the groundwater system where it can be transformed by micro-bacterial action over longer time periods. It is evident that one should refrain from putting these chemicals on the dry earth where they are poorly disseminated. It is also evident that each groundwater system has a certain capacity for such chemicals and that this capacity should be measured or predicted before one begins intensive or prolonged chemical dumping.

6.2 NUCLEAR WASTE DISPOSAL

A topic of current interest and anxiety is nuclear waste disposal. This is a problem controlled by water–rock and water–clay interactions. The problem is complex, has generated much controversy and a large body of competent scholarly research. Unfortunately the problem has not always been clearly stated and, more importantly, it has had its basic premises changed during the elaboration of the studies made to determine the important parameters. A detailed description of the problems of nuclear waste might seem beyond the scope of a book dealing with clays but the clay point of view has occurred at different stages of the scenarios of the waste problem, in its various forms. A reasonably descriptive background might be instructive to a potential clay mineralogist.

The basic problem in nuclear waste containment is one of kinetics. One wishes to keep the chemicals which are noxious to living things at a safe distance for a reasonable period of time. The greater the timespan of their isolation, the less noxious they become, through the agent of spontaneous radioactive decay (zero-order kinetics, which do not depend upon the concentration of the species considered).

Other kinetic factors are involved in the nuclear waste problem which are a function of time, of course, but also of temperature. These kinetics involve the transformation of the container materials and various barriers placed around the radioactive elements. The higher the temperature, the faster these reactions occur.

In the problems of agricultural or industrial waste in surface conditions, the problem of temperature could be ignored because all reactions take place at or near the surface of the solid earth where the temperature is modulated by the ambient effects of atmosphere. In problems of nuclear waste, the temperature is a critical factor.

6.2.1 The tree of choices in waste disposal

Waste disposal is a typical situation where scientific and engineering considerations come into play. If one considers the problem as a series of steps in time or as a sequence of time-related events which are conditioned by the industrial process, each step in the series demands a decision which determines the subsequent sequence of choices and operations. At each point a decision determines the whole sequence of choices and operations which will follow. This creates a sort of tree of processes or treatments whose branches are independent of and do not cross one another. One choice leads to another series of choices but they do not come round to the same solutions. Therefore, a geologist or natural scientist can be asked what is the best situation for nuclear waste, but his response is conditioned by the choices made for engineering reasons in the initial stages of the preparation of the waste material. A change in the preparation choices engenders a complete change in the choice of the geological setting. The same is true for a clay mineralogist. If certain choices are made in the conditioning process and the place of storage, the conditions under which the clays will be utilized will be completely different and the choice of clay materials will be correspondingly different.

Fundamentals of the problem

Let us look at the basic situation of nuclear waste disposal. Essentially, one wishes to isolate a certain series of radioactive (highly poisonous) elements from the biosphere. The major vector of introduction of radio-elements into the biosphere is in aqueous solution, groundwater or lake and ocean water bodies.

The radio-elements gradually lose their lethal characteristics as a result spontaneous decay. However, as they do so most of them release a non-negligible quantity of heat. This heat changes the environment around them *vis-à-vis* the natural state. An increase of heat increases the rate of reactions concerning the containers and barriers which have been devised to isolate the waste material.

Since the nuclear reactors already exist, and produce electricity at a relatively high cost, nobody wants to add greatly to this cost. Radioactive waste management must be accomplished with a minimum of cost.

These then are the basic problems of nuclear waste: time, heat and money.

The thermal problem

Thus, in any instance, when nuclear waste is disposed of somewhere it causes a thermal disturbance in its surroundings. Heat is produced at a rate roughly related to the radioactive decay of the unstable elements in the waste:

$$dx/dt = k$$

where the change in the concentration of the radio-element is a function of the decay constant k. The great mass of the heat is produced in the early stages of the decay.

One can diminish this initially high thermal disturbance in two ways: by leaving the substances in a holding area while they dissipate the greater part of their potential heat into the atmosphere (usually around 40–50 years for current uranium-based nuclear fuels); and by dispersing the elements into a large quantity of inert material so that the increment of heat is spread over a large volume and more easily dissipated by the ambient environment. Most organizations treating the nuclear waste problem have chosen to use the first method to eliminate the initial thermal problems. This solution also has the advantage of postponing the disposal decision by several decades.

Concentration of the remaining, thermally active, material then begins the series of choice and branch options: concentrated or dilute disposal techniques which mean high or low thermal flux in the initial, 'near-field' area of the waste depository. Whatever the choice, hot or warm, the next step is the isolation of the waste material in a more or less final repository.

Storage sites

Three types of storage site have been considered: surface; submarine sediment burial; and continental or land burial. All must be able to sustain an isolation period of up to a million years or so, depending upon the agency emitting the safety criteria. Here it is important to remember that the assumed period of development of the species of *Homo sapiens* to its present form is of the order of 100 000 years. Thus the projected protection period is much greater than that of evolution of sophisticated primates!

Surface repositories are considered to be unsafe because they would need to be controlled and kept isolated from biological interaction for periods of over a million years. It is considered, rightly, that such a task is beyond the capacity of present-day governments and institutions. Such repositories would be of reasonably low cost, however.

Submarine burial in unconsolidated clays is rather frowned upon because it can only be safely operated in very deep-sea sediments, putting the waste far from the normal biosphere. The thick layers of deep-sea sediments are normally found at water depths of 4–5 km. These are outside the territorial waters of almost all countries and hence belong to the entire planet. Burial in deep-sea sediments is essentially burial in everyone's back yard. Such a practice is generally considered unethical.

A second drawback to deep-sea disposal is that the waste, once buried, is irretrievable. No further investigation concerning its state and toxicity can be made. Such irreversible practices are also considered unwise. This method is, however, the lowest in cost, accomplished essentially by throwing protected capsules into the deep ocean and hoping for the best.

The third and most likely repository site is land burial. Here one can control access to the waste, one can control and monitor the interaction of water with the waste, and one has national jurisdiction over the disposal site. However, this method is the most expensive of the three alternatives.

Land repositories will be considered here as the most likely scenario of nuclear waste disposal.

6.2.2 High-energy waste

Nuclear waste can be classified according to the intensity and length of the decay process. Low-activity (short-lived and low-energy) material is buried near the surface or in surface protected sites. Few precautions concerning aqueous interaction are needed given the short residence time. Metal containers are stocked in concrete bins. Inspection is easy and corrosion is unlikely. Medium-activity waste is stored in much the same manner but at greater depth, tens to hundreds of metres.

High-energy waste is much more of a problem as the isolation time must approach a million years or so. Also this material produces large amounts of heat and the ionizing radiation released has a high penetrative power. Complete isolation is of the utmost importance. In such cases much thought and experiment has gone into assessing the different problems which will be encountered in efforts to isolate such material. Basically nuclear waste research deals with high-energy waste material, especially for those concerned with the environment.

Storage configuration

In all cases of proposed burial techniques, the situation is quite similar in concept, and only the design is different. The radioactive elements are confined in a concentrated solid form, either in glass or in synthetic rock, an assemblage of synthetic minerals. This material is considered to be relatively unstable in the presence of water and hence must be isolated from the enclosing rock environment. The first, the canister protecting the radioactive waste is usually a cladding of an inert metal such as copper or zirconium. The second barrier is some sort of mechanical system such as concrete used to fix the position of the waste package. The third barrier is

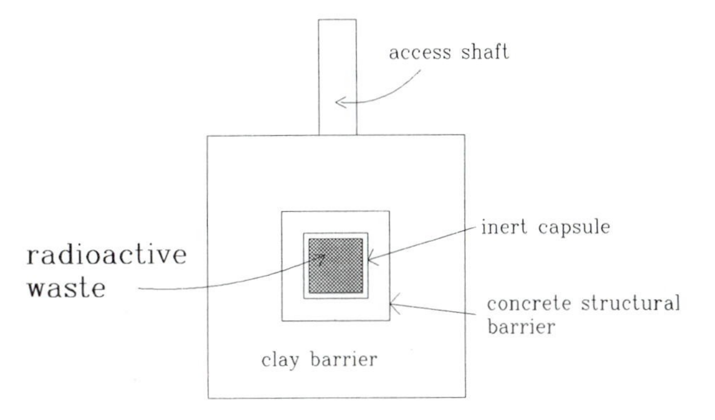

Fig. 6.11 General nuclear repository site for high activity waste, showing the different structural barriers to migration of the waste material. This device is situated in such a way as to give access to the waste at any time according to existing practice in most waste-producing countries at the present time.

that of compacted clay or zeolites used to absorb possible noxious elements which would have escaped the first barriers. The fourth barrier is the enclosing rock system which will be chosen for specific qualities which oppose migration of solutions. These relations are shown in Fig. 6.11.

Clay barriers

The third barrier, that of compacted, dried clays, is common to most repository strategies and has *two functions*. The dried clay will expand greatly as it absorbs water between its layers when attacked by aqueous solutions. This makes a mechanical seal between the solid barrier holding the waste canister and the rock walls of the enclosing geological formation. This physical seal creates a highly impermeable zone which should greatly impede fluid flow towards the waste canisters.

At the same time the clays wetted by aqueous solution will act as a reservoir of exchange sites to gather in the stray radioactive elements in the aqueous solution should they have escaped from their prison. This cation exchange action will, of course, be subject to the normal re-exchange process between solution and clay sites described in the demonstration of pollutants in soil profiles. That is to say, they can accept or release ions depending upon the concentration of the ions in solution relative to that on the clays. If an initial 'contaminated' solution saturates the sites, a fixed proportion of ions will nevertheless remain in the solution depending upon the partition coefficient between solution and clay. Thus the clays take in a major part of the radioactive elements in solution but not all of them. If subsequent ion-free solutions continue to come into contact with the saturated clay, there will be a transfer of ions to the solution. These ions will be dispersed away from the burial site into the 'far field' area. The tendency in ion exchange is for dispersal from high concentrations to more widespread low concentrations.

The third, clay barrier is in fact a retardant to dispersal of radioactive substances. The clays do not fix the ions permanently. This realization is fundamental to all of the chemical barrier strategies.

The long-term efficiency of the near-field clay barrier will depend upon its thermal stability, for if the waste heats up the environment, the smectitic clays are likely to change their crystalline state and became micaceous as in diagenesis. Such an effect is currently the subject of much research. The kinetics of clay transformation must be weighed against the eventual number of exchange sites remaining as a function of time and temperature. The higher the temperature, the faster the reaction will proceed and the more rapidly the exchange sites will vanish. At lower temperatures the clays are more stable but they are closer to the surface of the earth from which they must be kept.

If deep burial is chosen, the concentration of the waste material must be reduced to avoid overheating of the clay barrier. This, of course, adds to the cost of the disposal site by increasing its volume at depth. Therefore, the depth of the burial site plays a role in the efficiency of the clay barrier.

The surrounding rock which holds or harbours the high-energy waste can be considered as a type of retardant system to dispersal of the radioactive elements into the biosphere. In most silicate rocks occurring near the surface, there are a large number of clays present. In shales, for instance, the natural retention capacity due to normal cation exchange is very great. Thus shales are considered as a potential disposal medium. In granites, otherwise homogeneous and impermeable, one finds clays concentrated along fissures and cracks. These fissures and cracks are the natural passageways for aqueous solutions which could come in contact with the waste material. As the solutions leave the waste site, waste-laden as they might become, they will follow the same pathways and be in contact with clay minerals for much of their eventual trajectory. These clays will, of course, exchange the normal inoffensive ions such as Na, Ca and K for radioactive ions such as strontium and uranium. If the burial site is several kilometres from the surface, there will be many exchange sites between the release site and the eventual contact with the biosphere.

The last potential barrier to migration of nuclear waste is perhaps the most efficient over a long period of time.

Time is of greatest importance. If the migration path is tortuous it will slow the migration of fluids, thus permitting them to lose their lethal characteristics as they are retarded. Of course, the various physical parameters in such a complicated system allow scientists much latitude in choosing different scenarios which they can model with computer technology as the system progresses in speed and precision.

6.3 GENERAL CONCLUSION

In a chapter like this few things can be said that have not already been said clearly enough to be self-evident. Therefore these concluding remarks here will be brief. The study of clays is interesting and complex, in part because much knowledge is only recent and has not been entirely assimilated into the general corpus of scientific knowledge. In fact the properties of clays are simply divided into swelling and non-swelling, which gives a fundamental difference in all types of behaviour. The use of clays is determined by their chemical and physical behaviours. The importance of clays to the activities of man and beast, and to the plants upon which they feed, is also largely determined by the physico-chemical properties of clays.

The interaction of clays with their environment is cardinal to the ecosystem. Most of the interaction between man and nature is filtered through a clay membrane which protects and, at the same time, threatens human activity which can produce negative effects on living matter. If one understands the fundamental properties of clays they can be used or manipulated to the great benefit of mankind. There is no reason to fear the clay part of the environment. There is no reason to abstain from all activity which has been so useful and essential to the human race just because some instances of neglect and poor understanding of the principles of clay mineralogy have caused an ecological mishap. Knowledge is power, but only in the hands of the wise.

Index